RELIGION AND CHEMISTRY

RELIGION AND CHEMISTRY;

OR,

PROOFS OF GOD'S PLAN IN THE ATMOSPHERE AND ITS ELEMENTS.

JOSIAH P. COOKE, JR.

GYAN

Published by

Gyan Publishing House
5, Ansari Road
Daryaganj, New Delhi-110002
Phone: 011-47034999, 9811692060
E-mail: books@gyanbooks.com

Distribution Network
gyanbooks.com
India, USA, Canada, UK, Australia, France

ISBN: 978-81-212-8909-2 (PB)
First Published, 1864

2nd Impression 2023

Printed at: Gyan Press, Delhi.

RELIGION AND CHEMISTRY

AUTHOR: JOSIAH P. COOKE

RELIGION AND CHEMISTRY;

OR,

PROOFS OF GOD'S PLAN IN THE ATMOSPHERE AND ITS ELEMENTS.

TEN LECTURES

DELIVERED AT THE BROOKLYN INSTITUTE, BROOKLYN, N. Y.

ON THE GRAHAM FOUNDATION.

BY

JOSIAH P. COOKE, JR.,

ERVING PROFESSOR OF CHEMISTRY AND MINERALOGY
IN HARVARD UNIVERSITY.

"Duo sunt quæ in cognitionem Dei ducunt; Creatura et Scriptura."
ST. AUGUSTINE.

NEW YORK:
CHARLES SCRIBNER, 124 GRAND STREET.
1864.

PREFACE.

THE Lectures now published were first delivered before the Brooklyn Institute on Sunday evenings of January and February, 1861, and the larger part of them were subsequently repeated, during the same winter, before the Lowell Institute in Boston, and before the Mechanics' Association of Lowell. The progress of science since that time has rendered necessary many additions, and in revising the Lectures for publication, the material has been thus so greatly increased that what was originally prepared and delivered as six Lectures is now distributed over ten. At the time when the Lectures were written, Mr. Darwin's book on the Origin of Species, then recently published, was exciting great attention, and was thought by many to have an injurious bearing on the argument for design. It was, therefore, made the chief aim of these Lectures to show that there is abundant evidence of design in the properties of the chemical elements alone, and hence that the great argument of Natural Theology rests upon a basis which no theories of organic development can shake. In illustrating his subject, the author has used freely all the materials at his command, and if, in any case, he has failed to give due acknowledgment, it has been because by long dwelling on the subject the thoughts of others have become blended with his own. He would here acknowledge his repeated indebtedness to Professor Guyot's work on "Earth and Man," to Professor Faraday's published Courses of Elementary Lectures, and to Professor Tyndall's Lectures on "Heat considered as a Mode of Motion." He would also express his especial obligations to the author of "In Memoriam," in whose verses he has discovered a truer appreciation of the difficulties which beset the questions discussed in this volume, than he has ever found in the philosophy of the schools.

CAMBRIDGE, May 3d, 1864.

CONTENTS.

GRAHAM LECTURES.

RELIGION AND CHEMISTRY.

LECTURE I.

STATEMENT OF THE CASE.—TESTIMONY OF THE ATMOSPHERE.

The relation of the Church to physical science.

THE time has been when the Christian Church was an active antagonist of physical science; when the whole hierarchy of Rome united to condemn its results and to resist its progress; when the immediate reward of great discoveries was obloquy and persecution. But all this has passed. The age of dogmatism has gone, and an age of general scepticism has succeeded. The power of traditional authority has given place to the power of ideas, and physical science, which before hardly dared to assert its birthright, and could even be forced to recant, on its knees, its demonstrated truths, has now become one of the rulers of society. By its rapid growth, by its conquests over brute matter, and by its wonderful revelations, it has deservedly gained the highest respect of man, while by multiplying and diffusing the comforts of life it has become his acknowledged friend. Every effort is now made to further its progress. Its great discoveries win the applause of nations, and its fortunate students are remembered when the princes and nobles of the earth are forgotten.

All this is well. But unfortunately, elated by his success, the stripling has been at times proud and arrogant, usurping authority not his due. Forgetting his early faith, he has approached with irreverent thoughts the holy temple of our religion, and, not content to worship in the outer court, has dared to penetrate into the very Holy of Holies, and apply his material tests even to the vessels of the altar. No wonder that the Church should become alarmed, that many of her best men, holding fast to the sacred dogmas of our religion as the only sure anchor of their faith in this world, and their sole ground of hope for the next, should join in a general cry against the whole tendency of science and its results.

But this is a great mistake. Judging of the real character of physical science from the pretensions of a few, and not possessing the power or opportunity of investigating for themselves, these good men are unnecessarily alarmed: the phantom they fear is purely of their own creation, and, could they but know the whole truth, they themselves would see that to ignore the well-established results of science, and to denounce its legitimate tendency, is a policy as short-sighted as it is illiberal and unchristian.

Fortunately, such fearful souls constitute but a small party in the Christian Church. There is a far nobler and more courageous faith than theirs, — a faith so strong in its convictions that it fears no criticism, however searching, and no scientific analysis, however rigorous it may be, — a faith which finds in the Bible, not a series of dead formulas, but a mass of living truth, — a faith which really believes that the God of nature is the God of grace, and that man was created in the image of this one and only God, — a faith which wells up from the depths of the soul, which speaks because it believes, which believes because it feels, — a faith whose sources are as hidden as those of the fountain, but whose reality is as living as the verdant landscape which the fountain waters.

The attitude of a brave Christian faith.

It is the men with a faith like this who are the really brave Christians. They are not alarmed at the apparent contradictions between science and revelation. By the very imperfections of their own faculties, which they so keenly feel, they have been taught that mysteries exist; nay, they find in these very mysteries the strongest bulwarks of their faith; for they feel, with Robert Hall, that "a religion without its mysteries would be a temple without its God." They are fully assured that our minds were framed after the likeness of their Divine original, in order that we, creatures of the dust though we are, might nevertheless in our feeble measure comprehend

God's workmanship and sympathize with his divine thoughts; and they reject as absurd the doctrine that man, thus created an intelligent and sympathizing observer of God's universe, should have been permitted, in the legitimate exercise of the very powers which God has given him, to build up a connected system of science in direct contradiction with those higher and spiritual truths which the Father has been mercifully pleased to reveal to his sinning children through his prophets and his Son.

May the foundation of these lectures ever stand as a monument of this brave Christian faith. He who provided in his last testament that the wisdom, power, and goodness of his God should at short intervals be illustrated in this place from the varied stores of knowledge which the investigator of nature might collect, could not have feared that there was any essential contradiction between science and revelation. On the contrary, he must have felt that science would become the great champion of the Christian faith; and may the result fully prove his wisdom and his foresight.

The whole material creation illustrates God's attributes.

It will be my object, in the present course of lectures, to aid in carrying out the noble purposes of the founder by developing before you the illustrations of the wisdom, goodness, and power of God, which have been discovered in the constitution of matter. The material of this universe, by which we are surrounded and in which we live and move, is full of the evidences of these attributes of the Creator. They crowd upon us from every side. Wherever we turn our eyes there we read them, on the blue dome of heaven, and on the gorgeous cloud-turrets of the western sky; they are inscribed on the rocky cliffs which record the memory of long-buried ages, and on the green sods which cover the last-made grave; they are heard in the crackling of the conflagration no less than in the roaring of the tempest, in the pattering of the rain-drop as well as in the rolling of the thunder.

"The heavens declare the glory of God, and "the firmament showeth his handywork. Day "unto day uttereth speech, and night unto night "showeth knowledge. There is no speech nor lan-"guage where their voice is not heard. Their line "is gone out through all the earth, and their words "to the end of the world."

It is the privilege and duty of the student of nature to interpret the characters in which these evidences are written, and the language which nature speaks; but the best translation gives only an inadequate idea even of a work of human genius, and far less can my imperfect rendering convey to you the full force of that impression which the devout student receives while studying the perfect works of God.

The illustrations of the attributes of God, which may be drawn from the constitution of matter, are conveniently divided into two classes,—first, those which appear in the adaptation of various means to a particular end, and, second, those which are to be found in the unity of plan according to which the whole frame of nature has been constructed. The first class are exhibited by the properties of matter, the second by the so-called physical laws and forces.

Two classes of illustrations.

There are probably no forms of matter in which, even with our present imperfect knowledge, a combination of means conspiring to some beneficent end cannot be traced; but such adaptations are more especially to be seen in the properties of those substances which are the most intimately associated with man, and on which his existence and happiness depend. We shall not, therefore, be obliged to search the recondite facts of science, or to investigate the rarities of nature, to find our illustrations of the Divine attributes; they lie all around us, in the air we breathe, in the water we drink, and in the coal we burn; and what more convincing evidence could we have of the beneficence of our Heavenly Father than the very fact that these familiar substances are of all others most vocal with his praise?

In following out, then, the order which seems to be so obviously indicated by the nature of the case, I shall ask you, in the first place, to study with me the physical condition of our atmosphere, and the properties of the various materials of which it consists; and I am

The plan of this course of lectures.

sure we shall not fail to find in one and all abundant evidence of the wisdom, goodness, and power of God. Having thus made you acquainted with some of the more important scientific facts required for my argument, I shall next direct your attention to those grander demonstrations of God's wisdom and power which appear in the great laws and forces, by which the whole material universe is upheld, and lastly an examination of the relative limits of scientific and religious thought will form an appropriate termination for the course.

The argument from special adaptations.

The argument from special adaptations which lies at the basis of all works on natural theology is admirably condensed by Dugald Stewart into two simple propositions. The one is, "that everything which begins to exist must have a cause"; the other, "that a combination of means conspiring to a particular end implies intelligence." To these might be added the two equally clear propositions stated by Dr. Reid: first, "that design may be traced from its effects"; second, "that there are evidences of design in the universe." I do not intend to discuss at length the logical validity of this argument, or the general value of analogical reasoning which it implies. Such discussions belong particularly to the province of metaphysics, and I willingly leave them to abler hands. It will be my chief object in these lectures to bring to your notice a few of the numberless indications of adaptation in the materials of our atmosphere, assuming for the present that these adaptations are evidences of design, and therefore evidences of the existence of a personal God, infinite in wisdom, absolute in power. When we have thus become acquainted with some of the facts on which the argument rests, we may then profitably consider the validity of the reasoning, at least so far as to weigh the objections, which modern materialism has urged against it.

Prejudice against natural theology.

I am well aware that great prejudice is entertained in many religious minds against the whole scheme of natural theology, originating in a vague idea that its study is in some measure unfavorable to the growth of Christian faith. That such an impression should be produced by a class of modern writers, who think to supersede Christianity by a sort of pantheistic naturalism, is not surprising; but certainly natural theology is no more chargeable with the presumption of such speculations than is Christianity with the excesses of Jesuitism. Let me not be misunderstood. I have no desire to press the argument of natural theology too far.

Natural theology no substitute for an historical faith.

For myself, I believe that the facts of human nature themselves all tend to prove that a divine revelation is the only legitimate basis for a system of religion, and that an historical faith based on a supernatural revelation is the only religion possible for imperfect humanity. Indeed, I am led to think we find evidence of the goodness of our Heavenly Father in the very circumstance that the foundations of all knowledge have been laid in such obscurity that no unaided human intellect can wholly dispel the cloud which hides the Creator from our sight,—

> "To feel, although no tongue can prove,
> That every cloud that spreads above
> And veileth love, itself is love."

This very obscurity humbles the pride of human learning, and raises its constant warning against that intellectual idolatry which would substitute its shallow philosophy for the simple truth as it is in Jesus. The Bible once received, science can furnish abundant illustrations of the attributes of the Being therein revealed; but even with all the illumination which has been the immediate or secondary result of Christianity, man is hopeless without its authority, and I would not give the slightest shadow of support to that irreverent presumption which, guided by what it calls the unaided light of nature, would construct a system of religion out of passions, intuitions, and I know not what absurdity.

But still it must be remembered that the Christian revelation does not prove the existence of God; on the contrary, it is based on a belief in his being that already exists in the mind of man. The Bible opens with this assumption. The first line asserts that

Natural theology the logical basis of Christianity.

"In the beginning *God* created the heaven and "the earth."

And the Hebrew name of God, Jehovah (*I am that I am*), is itself a declaration of his self-existent being.

With all men a belief in some Almighty Power overshadowing their being grows up spontaneously in the heart, they know not how; but the educated and the intelligent seek further to find its logical grounds in the evidences of nature.

Here, then, is the first great office of natural theology. It furnishes the logical basis on which the whole scheme of revealed religion given us in the Bible rests.

I have no desire to over-estimate the importance of my subject. For myself, I believe, with Paley and the other eminent writers of the same class, that the fundamental truths of our religion can be inferred from the constitution of the human mind and from the course of nature with as much certainty as analogical reasoning can ever give. But still I know that the evidence is not demonstrative and not likely to convince the sceptic; for in the last analysis it rests on certain assumptions which he will not admit. And it is in vain to urge that these assumptions are really intuitive truths and tacitly admitted by the whole human race; for he easily replies, that they are not intuitive to his mind.

Nevertheless, the evidences of God in nature are the only proof we have or can have of his existence, and they are therefore the only logical basis of the Christian revelation. Nature and revelation are parts of one and the same system, and, however much our prejudices may obscure the fact, Christianity rests on natural religion, and cannot be logically defended if the authority of the last is denied.*

Natural theology confirms and illustrates the truths of revelation.

But however great the value of natural theology, considered as the basis on which revelation rests, this is not its only or most important office. In the present age of the world it confers a still more inestimable benefit on mankind by confirming, illustrating, and enforcing the admitted truths of revelation.

If it be asked of what value are further illustrations of admitted truths, I answer, that there is an important class of nominal Christians who are more open to impressions from the study of nature than to direct appeals to the heart. It is true that the great mass of mankind must be Christianized, if at all, through the affections and by the hard discipline of sorrow; but there are some who, not yet tried in the fiery furnace of affliction, have first felt their Father's hand and recognized his love while contemplating his works. I do not say that persons so touched are already Christians, but I do say that the first step has been taken, and that is a great deal. It may require many years of sad experience and many a bitter pang of disappointment before they come to kneel humbly at their Saviour's feet; but, like the great Apostle, they will always look back to the time when the Divine presence first visibly

* See a recent work by the Rev. A. P. Peabody, D. D., entitled, "Christianity the Religion of Nature," for a development of this idea.

shone before them as the turning period of their life.

While, therefore, I should be the first to condemn that hollow naturalism which would substitute a

Importance of this office.

system of natural theism for the simple doctrines of the Bible, I must also deprecate that prejudice which prevents many clergymen, through fear of this tendency of the age, from availing themselves of the aid of science in enforcing the fundamental truths of our religion. I assure them they thus neglect a most important means of influence over educated and thinking men, — a means of influence always important, but never more so than in an age which is marked by its cultivation of practical science, and in a country where so large a portion of the active energy of the community has taken this practical direction. The danger of our time is not so much a philosophical scepticism as a practical materialism. The fear is, not that men should reason themselves into unbelief, but that, spending their whole lives in developing the powers of nature, they should practically worship the dead matter, rather than the living God. If, however, you can make such persons feel that the material is but a form of the spiritual, and that in fact the spiritual is nowhere more manifest than in those very laws and forces which they so much idolize, you will not change, it is true, the tendency of the age, but you will ennoble and sanctify it. The whole material universe will become transfigured, and nature will no longer be seen as a wonderful mechanical application of blind forces, but as a living embodiment of the Eternal One. Nature-worship may continue, but it will have lost its idolatry; for it will be no longer the machine that is worshipped, but that same Living Spirit which spoke in tones of thunder from the clouds of Sinai and in accents of mercy at the baptism of Christ.

I know it is said that nature conceals rather than reveals God, and in a certain sense it is undoubtedly true that He is hidden from us behind the veil of his works; but since it is permitted to man by the exercise of his intelligence to lift in part this veil, it is certainly the duty, as it should be the privilege, of the ministers of religion to show forth the unspeakable glory which lies behind these material forms.

Example of Christ.

But why multiply arguments when we have the authority of the Great Teacher himself, who frequently appealed to nature to illustrate and enforce the divine truths which he came on earth to reveal? We have indeed the whole summary of Natural Theology in His simple words. "Wherefore, if God so clothe the grass of the "field, which to-day is, and to-morrow is cast into "the oven, shall he not much more clothe you, O "ye of little faith?"

With, then, such authority as this, let us not despise the beginnings because they are not the end, or undervalue the means by which many a noble soul has been led to the foot of the Cross.

Without seeking, therefore, to vindicate further the claims of my subject, I will at once enter upon the plan already proposed for this course of lectures, and will first ask your attention to the illustrations of the wisdom, goodness, and power of God, which may be discovered in the constitution of our atmosphere. In endeavoring to carry out this plan, I shall require all your indulgence and all your kind forbearance. From the very nature of the case, it will be necessary to start from first principles, and much of the way we are to travel together will be uninteresting and dull. If, however, the path shall lead us to the summit of that holy mountain from which we can gain a clearer vision of spiritual things, we shall soon forget the toil and difficulty of the ascent. We have no extravagant expectations of the result. We do not hope to convince the sceptic, or to arouse the indifferent from their practical unbelief. Our only hope is, — and this we entertain in all humility, — that, by pointing out a few of the footprints of the Creator which lie thickly along our daily path, we may encourage some earnest student toiling forward on his journey of life. May God grant to us all the richest blessings of His grace; for though man may plant and water, He only giveth the increase.

The illustrations of the attributes of God presented to us by the atmosphere are especially manifest in those adaptations of properties by which it has been made to subserve the welfare and happiness of mankind. Here, however, as always in the study of nature, we must be careful to avoid the error of considering man as the sole end of creation, and of interpreting all phenomena with reference to him alone. The material universe is the manifestation of one grand creative thought, as comprehensive in the diversity of the parts, as it is grand in the unity of the whole. These parts have been so wondrously joined and skilfully wrought together, that each is linked with each, and one with all. In this Divine economy nothing is wanting, nothing is superfluous, and what seems to our feeble vision least important is as essential to complete the unity of the plan as our own glorious manhood:

Adaptations in the atmosphere to be studied with especial reference to man.

> "Nothing useless is or low,
> Each thing in its place is best,
> And what seems but idle show
> Strengthens and supports the rest."

Amidst all this wonderful variety in unity, man stands the culminating glory of the whole. Made in the image of his Creator, and but "a little lower than the angels," he has been intrusted with dominion and power over all the brute matter which surrounds him. Through the long ages of geological history the earth was preparing for his dwelling, and in the earliest forms of animal life his coming was prefigured and foretold. It will be natural, therefore, to consider the adaptations of the atmosphere with special reference to him; and this we may do legitimately, without losing sight of the grand idea which underlies the whole, and of which man is only the nobler part.

Description of the atmosphere.

The atmosphere is a vast ocean of aeriform matter, enveloping the earth like a mantle, and rising to the height of many miles above our heads, but constantly diminishing in density as the elevation increases. At the height of about three miles and a half (3.43) the density is only one half as great as at the level of the sea; and at the height of forty miles it is less than in the exhausted receiver of the best air-pumps. How much higher than this the atmosphere extends, it is impossible to determine with accuracy. In this ocean of air all bodies on the surface of the globe are immersed. It is so subtile that it penetrates into the minute pores of matter, and fills the cavities of all organized being. It is the medium in which all vital processes both of plants and animals take place, and in which all human activity has its seat. Let us see now with what wisdom its properties have been adapted to the important ends which it is appointed to subserve.

Physical state of the atmosphere.

Consider, in the first place, the physical state of the atmosphere, its very aeriform condition. This air is as truly matter as the solid planks on which we are treading, or the granite rocks on which this building rests. It is far less dense, it is true, but then it has all the essential properties of matter. It fills space. It resists with an ever-increasing force all attempts to condense it; and, moreover, it has weight. But how different in condition from the solid rock!—so different that to the uneducated it hardly seems material; and in our common language we speak of a space which is filled only with air as empty. Its particles are endowed with such perfect freedom of motion, and yield so readily to the slightest pressure, that we move through it without feeling its presence. It is firm enough to support the wings of the lark as he mounts the sky, and yet so yielding as not to detain the tiniest insect in its rapid flight.

Two opposing forces acting in the atmosphere.

The physical condition of the atmosphere will still further excite our admiration, when we consider the wonderful play of forces by which it is upheld. It may not be known to you all that upon this mass of air, outwardly so calm and passive, there are constantly acting two mighty forces,—the force of gravitation and the force of heat. In virtue of the force of heat the particles of the atmosphere mutually repel each other, and the whole mass, like a great bent spring, tends to break from its confinement and to expand into the surrounding space; but this it cannot do, for by the power of gravitation it is held with a firm grasp to the surface of the globe. Were this grasp for a moment relaxed, the atmosphere would dash off with explosive violence and be lost in the immensity which surrounds us. How great the force is which is required to restrain the expansive tendency of the atmosphere few persons have an adequate conception, because the two opposing forces are so perfectly balanced that we are obliged to call in the aid of experiment in order to render their effects evident. So true is this, that the world never even dreamed of their existence until within two hundred years, and the story of the discovery is one of the most remarkable in the history of inductive philosophy. This story is well known; but as it is short, and teaches us an important truth, you will pardon its repetition.

"Nature abhors a vacuum."

Every one who has seen a common pump is familiar with the fact that it is the pressure of the air which causes the water to rise in the suction-pipe, and this suction is one manifestation of that force by which the atmosphere is held so firmly to the surface of the globe. The pump, however, was used long before the discovery of the pressure of the atmosphere, and its action was explained by a principle which seemed perfectly satisfactory then, but which sounds strange enough to modern ears. The principle appears first to have originated with the Aristotelians, and was expressed in the phrase, "Nature abhors a vacuum." These ancient philosophers noticed that space was always filled with some material substance, and that the moment a solid body was removed air or water always rushed in to fill the empty space. Hence they concluded that it was a universal law of nature that space could not exist unoccupied by matter, and the phrase just quoted was merely their figurative expression of this philosophical idea. When, for example, the piston of a common pump was drawn up, the rise of the water was explained by declaring, that, as from the nature of things a vacuum could not exist, the water necessarily filled the space deserted by the piston.

Florentine pump-makers and Galileo.

This physical dogma served the purpose of natural philosophy for two thousand years, and it was not until the seventeenth century that men discovered any limit to nature's abhorrence of a vacuum. Near the middle of the century some engineers were employed by the Duke of Tuscany to sink a well in the neighborhood of Florence to an unusual depth. They finished their work, but on adjusting the pump they found to their surprise that it would not work. With all their efforts the water would rise only a little more than thirty feet, and by no ingenuity or skill could they raise it an inch higher. More disgusted with nature than nature was with the vacuum in their pump, they applied to Galileo, then an old man, living in his villa on the brow of Fiesole. He could not aid them, but he is said to have replied, half in jest, half in earnest, that nature did not abhor a vacuum above ten metres. Had this incident occurred earlier in his career, Galileo would undoubtedly have added to the other jewels of his crown a brighter gem than all, but now the vigor of his manhood was spent; he had done his work, and, worn out by the persecution of a bigoted priesthood, he was peacefully resting from his life's labor, and calmly awaiting the close.

But the key which the incident had furnished was not lost. It passed into able hands, and it was the fortune of Torricelli, Galileo's best pupil, to unlock the secret. This young Italian philosopher, whose clear intellect had been trained in the mechanical philosophy of his great master, saw at once that a column of water thirty-three feet high, and no higher, could not be sustained in a cylindrical tube by a mere metaphysical abstraction.

This effect, he said, must be the result of some mechanical force equivalent to the weight of the mass of water sustained. It was not difficult to prove the correctness of this reasoning, for it was evident that if a column of *water* was sustained at the height of thirty-three feet in the suction-pipe of a pump by a constant force, the same force could only sustain a column of a *heavier liquid* at a proportionally less height. So Torricelli tried mercury, a liquid thirteen and a half times heavier than water, and the result was as he had anticipated. The force which raised the column of water thirty-three feet could only raise a column of mercury to the height of thirty inches, which is thirteen and a half times less than thirty-three feet. Torricelli did not, however, make this experiment with a pump, but with an apparatus of his own, much simpler and equally effective.

He took a long glass tube, open at one end, filled it with mercury, and, having closed the opening with his thumb, inverted the tube, and plunged the open end in a basin of mercury; on removing his thumb the mercury, instead of remaining in the tube, and thus satisfying nature's abhorrence of a vacuum, fell, as he expected, and, after a few oscillations, came to rest at a height of about thirty inches above the level of the mercury in the basin. The correctness of his induction having been thus verified, Torricelli at once concluded that it must be the pressure of the air which sustained both the water in the pump and the mercury in his tube.

Experiment of Torricelli.

This experiment excited a great sensation in Europe; but, as might naturally have been expected, the old physical dogma was not easily laid aside, and Torricelli did not live to see his opinion generally received. It was left to the celebrated Blaise Pascal to convince the world that Torricelli was right, and this he did by one of those master strokes of genius which at once silence controversy.

Demonstration of Pascal.

"If," said Pascal, "it be really the weight of the atmosphere under which we live that supports the column of mercury in Torricelli's tube, we shall find, by transporting this tube upwards in the atmosphere, that in proportion as it leaves below it more and more of the air, and has consequently less and less above it, there will be a less column sustained in the tube, inasmuch as the weight of the air above the tube, which is declared by Torricelli to be the force which sustains it, will be diminished by the increased elevation of the tube."

Accordingly Pascal carried the tube to the top of a church-steeple in Paris, and observed that the mercury fell slightly, but not satisfied with this result, he wrote to his brother-in-law, who lived near the high mountain of Puy de Dôme, in Auvergne, to make the experiment there, where the result would be more decisive.

"You see," he writes, "that if it happens that the height of the mercury at the top of the hill be less than at the bottom, (which I have many reasons to believe, though all those who have thought about it are of a different opinion,) it will follow that the weight and pressure of the air are the sole cause of this suspension, and not the horror of a vacuum; since it is very certain that there is more air to weigh on it at the bottom than at the top; while we cannot say that nature abhors a vacuum at the foot of a mountain more than on its summit." M. Perrier, Pascal's correspondent, made the observation as he desired, and found a difference of about three inches, "which," as he replies, "ravished us with admiration and astonishment."

Thus it was that man first learned to recognize the existence of that power, which retains the atmosphere on the surface of the globe, and the history of the discovery should humble our intellectual pride and teach us to hold our knowledge with reverence and humility. This old scientific dogma of the seventeenth century never fails to excite a smile, and we are inclined to wonder how man could ever have believed what now appears so absurd; but if, like an antiquary, we imbue our minds with the spirit of that age, it will be seen, not only that the dogma was not essentially absurd, but also that the philosophical idea, clothed in those quaint terms, appeared to the scientific men of the period as truly a legitimate induction from observed facts, as the law of gravitation seems to us. And the induction was legitimate; but since the known facts did not cover the whole ground, they gave only a very partial truth. The Grand Duke's pump was the first failing case, and proved, not that the old principle was absolutely false, but only that its application was very limited.

The old dogma a partial truth.

Thanks to Galileo, Torricelli, Pascal, and Newton, — noble line of genius, — nature's abhorrence of a vacuum gave place to the law of gravitation, and two centuries of unparalleled scientific activity have only served to confirm the truth, and extend the domain of Newton's grand generalization; but even after this signal triumph, who now feels fully assured that the law of gravitation may not find its failing case? and when, two centuries hence, the future historian comes to write the history of inductive philosophy, who can feel certain that Aristotle's dogma and Newton's law may not both be condescendingly noticed among the partial truths which served the purposes of science in its infancy and childhood?

Let me not be understood to imply a belief that man cannot attain to any absolute scientific truth, for I believe that he can, and I feel that every great generalization brings him a step nearer to the promised goal; but I wish here at the outset most strongly to impress the distinction between the undoubted facts of science, and the laws and principles which have grown up around them, and by which they have been embodied in our systems of philosophy, — the distinction, in a word, between the observed phenomena of nature, and man's interpretation of the phenomena.

Distinction between fact and theory.

This distinction, so obvious when stated, is too often forgotten, and is necessarily overlooked in our scientific text-books. It is the sole aim of these elementary treatises to teach the present state of knowledge, and they would fail in their object if they attempted by a critical analysis to separate the phenomena from the laws or systems by which alone the facts of nature are correlated and rendered intelligible. But although while studying science itself, we may for the time waive the distinction between fact and theory, the moment we come to compare the results of science with the eternal truths of religion, the distinction here enforced becomes of paramount importance, and it must be our chief aim to separate that which is absolute and eternal truth from that which, even in its highest development, is the result of human thought, and, like all things human, subject to limitations and liable to change.

Had this distinction been always borne in mind, the controversies between the philosophers and the churchmen would have been less bitter and more fruitful in truth; the philosophers would have been willing to waive their theories, and the churchmen would have been led to respect the results of science, and conform their theology to the indisputable truths which God has been pleased to reveal through nature no less plainly than in his written word; and if the trite anecdote of Galileo and the pump-makers will serve to impress the distinction on our minds, this digression will not have been made in vain.

Pressure of the atmosphere.

You must all have recognized in Torricelli's tube our modern barometer. By means of this well-known instrument we can readily estimate the pressure of the atmosphere, and determine the amount in our human standards of measurement. It can be readily proved that the pressure of the atmosphere is about fifteen pounds on every square inch of the earth's surface, and if, starting from this well-known fact, you calculate the amount of pressure on any extended surface, you will be astonished at the result. For example, the pressure exerted by the atmosphere on the area on which this building stands is vastly greater than the whole weight of the building itself. The pressure on a man of ordinary stature is about sixteen tons; that on one square mile of surface is equal to over twenty-six million tons.

How great, then, must be the pressure on the whole surface of the globe, or, what is the same thing, how great is the intensity of that ever-acting power, which holds the atmosphere in its appointed place! It would not be difficult to calculate the amount and to express it in numbers; but these numbers would convey to you no definite idea, for our minds are incapable of forming an adequate conception of such immensity. The attempt to grasp it only exposes our weakness, and yet this force, immense as it is, is so delicately balanced by the sweet influences of the sunbeam, that it does not so much as shake the aspen-leaf or break the gossamer. If we believe no more than this, that the world was once created by God, what must be the power and wisdom of a being who could appoint these mighty forces and adjust them with such perfect precision! But if we also believe that these forces are direct emanations of Divine Power, — that it is God himself who with His own right hand holds the atmosphere in its place, and appoints its bounds, — then all nature assumes a more glorious aspect, and we feel that we are indeed surrounded by the Divine Presence. Yet this force, which we find so far beyond our powers of conception, is but a secondary phase of that immeasurably greater power which "brings forth Mazzaroth in his season, and guides Arcturus with his sons." How futile all attempts to measure Divine power! We select some one of the feeble forces acting around us, and succeed in reducing its value to our human standards of comparison, and expressing this value in numbers; but the numbers, when obtained, are beyond our grasp, and we find that we have merely mounted to a little higher platform, from which we discover numberless other forces immeasurably greater than the first. Something, however, has been gained. We have attained to the idea of the infinite; and to thoroughly apprehend the existence of the infinite, is to take the first step towards recognizing the existence of a God.

The idea of the infinite.

I know it will be said that man cannot comprehend the infinite, and if by this statement it is only meant to affirm the declaration of the Bible, that man cannot "find out the Almighty unto perfection," not even the most visionary dreamer would question the position. But there is a class of philosophers at the present day who think to enforce the authority of revelation by maintaining the doctrine that man can know absolutely nothing of the infinite,—nothing more than he now knows of the facts or principles of science to be hereafter discovered; that, indeed, the very term infinite implies a negation of all cognizable qualities.

To me, this position seems fatal to the very cause it is intended to defend, and surrenders all the approaches of the citadel to the infidel. For if there is in man no possibility of apprehending the infinite, even to the smallest degree, I can see nothing to which revelation can appeal. He has then no power to distinguish between the Divine and the human.

But it is not so. Revelation implies, and all experience shows, that man can recognize the presence of the infinite by attributes as clear and unmistakable as those which mark the presence of the finite matter around him. He may not be able to comprehend a single attribute of the infinite in its essence, but as the mathematician, dealing with infinitesimal quantities, which he cannot fully understand, arrives at truths of the material world with all the certainty of demonstration, so the mental philosopher may attain to moral truths in regard to the Infinite Being, although the very terms he employs may be veiled in impenetrable mystery.

And what is the true human conception of the infinite? It is not merely something which we feel to be very great indeed, but it is something which we feel surpasses our utmost conceptions of the great, — something which, let us account it as great as we please, yet, wherever the inability of our mental power fixes the limit of our conception, will still be felt to be greater than the greatest. We cannot gaze into the heavens without awe; we cannot examine the wonders of the dew-drop without reverence; we cannot look into our own souls without trembling. It is the same invisible Presence everywhere, and however long false philosophy may conceal the vision, or material cares and pleasures blind the senses, when man once recognizes its existence he instinctively worships and adores.

The evidence of design which we are now studying is still further enforced when we consider how exactly the density of the atmosphere has been adjusted to the human organism. By density is meant, I need not state, the quantity of matter which the atmosphere contains in a given volume; for example, in a cubic yard. This quantity is capable of exact measurement, and although to a certain extent variable, it is constant in the same place, under the same conditions of temperature and pressure.

Density of the atmosphere.

In this latitude, at the level of the sea, one cubic yard of the atmosphere when dry, and under the normal conditions of temperature and pressure, contains about two pounds of air, and this weight is the measure of its density. Now we find that the organization of plants and animals, including man, has been most skilfully adjusted to the density of the air, and many illustrations of this adaptation will be met with as we proceed. But accepting the fact for the present as universally conceded, let us consider the conditions on which this exact adaptation of the air to our physical organization rests.

The density of the atmosphere may be said to depend upon four conditions: first, on the inherent nature of the substance which we call air itself; secondly, on the intensity of gravity; thirdly, on the total quantity of air on the globe; and, lastly, on the temperature. The influence of the first condition is not understood, but that of the last three we can readily trace. If the intensity of the force of gravity at the surface of the earth were to change, other circumstances remaining the same, the density of the atmosphere would change in the same proportion. Thus, for example, if the intensity of gravity on the earth were as great as it is on the surface of the sun, the density of the atmosphere would be twenty-eight times as great as at present; or if this intensity were reduced to that which exists on the surface of the moon, the density would be diminished to one sixth of the existing density.

But, assuming that the intensity of the force of gravity on the surface of the earth remained constant, precisely the same effect would result from any variation in the total quantity of the atmosphere. Were the whole amount of air on the earth increased or diminished, the density of the atmosphere at its surface would also be increased or diminished in the same proportion. Still further, assuming that, while the intensity of gravity and the mass of the atmosphere remained fixed, the temperature were changed, then also the density of the atmosphere would vary, and by a quantity which can be easily determined. By accurate experiments it has been ascertained that an elevation of temperature equivalent to about five hundred degrees of our Fahrenheit thermometer would reduce the density to one half, and, on the other hand, that a reduction of temperature would increase the density in the same proportion.

Consider next what these relations imply. Reflect that the intensity of the force of gravity depends upon the mass of the earth. Remember that the mean temperature depends upon the distance of the earth from the sun, and you will see that not only the actual size of the earth, but also its distance from the sun, and the quantity of air on its surface, were all necessary conditions, in order that the atmosphere should have its present density, and thus become the fit abode for man. If any one of these conditions had been neglected, the result would not have been attained, and man, as he exists, could not have lived on this globe.

Evidence of unity in the design.

It must then have been He "who hath meted "out heaven with the span, and comprehended "the dust of the earth in a measure, and weighed "the mountains in scales, and the hills in a bal-"ance," who "formed man of the dust of the "ground, and breathed into his nostrils the breath "of life."

The unity of the design implies the oneness of the designer, and, although the adaptations just considered may not exclude every possible atheistic theory of cosmogony, yet they show conclusively that, if there is design anywhere, there is design everywhere ; *if there is design in the least, there is design also in the greatest, and design in the atom may thus confirm the evidence of design in man.*

LECTURE II.

TESTIMONY OF THE ATMOSPHERE. — *Concluded.*

Relation of the atmosphere to light.

DURING a recent journey in Switzerland, at the close of a delightful summer's day, in the early part of July, I arrived at Interlachen, in company with a number of fellow-travellers. We had been sailing on the beautiful lake of Brienz, and some minutes before we reached our destination the sun had set, and the mountains had already cast their long shadows across the lake. Early in the afternoon the clouds had settled on the nearer hills, and we had been disappointed at not obtaining a view of the distant summits of the Bernese Oberland; but suddenly, as the boat neared the shore, the magnificent peak of the Jungfrau appeared from behind the veil of clouds, clothed in her white mantle of everlasting snow, and bathed with a flood of rosy light. The effect thus heightened by the contrast was grand beyond description, and as beautiful as it was grand. It seemed like a vision of the Heavenly Kingdom, — as if the glory of God had rested on the mountain. The scene completely filled the soul, and the heart overflowed with gratitude for the blessing it enjoyed. It was felt to be one of the great privileges of a lifetime, and his would have been a dull understanding, and a duller heart, which did not recognize the Giver in the gift. The view so riveted the attention that we hardly noticed our arrival, and as we walked to the hotel we watched the successive shades of crimson and purple as they flitted up the mountain, until the last blended in the gray of the twilight.

It may not be permitted to many to behold the Jungfrau blushing before her retiring lord, but all have witnessed the same effect on even a grander scale, when the white clouds, piled up on the western horizon like vast mountain chains, become, at evening, resplendent with the rays of the setting sun; and many have watched their varying tints of gold and purple, until at last their ghostly forms vanished in the dusk of the evening, and the stars came out to take up with their measured twinkling the silent song of praise. Perhaps, also, there may be some who, after anxiously watching through the night, have felt their hearts strengthened and their hopes revived when the blush of morning reassured them of their Father's providence, and all nature smiled in the floods of returning light.

All these glorious visions, all this beauty, and all the pure emotions of our hearts which they excite, we owe, my friends, to the skill with which the physical qualities of the atmosphere have been adjusted to the wants of our physical and moral natures, and they all thus become the silent witnesses not only of the wisdom, but also of the goodness of our God.

We have already, in the first lecture, discussed some of the adaptations of the physical condition of the atmosphere to the purposes which it subserves on the globe, and I wish this evening to develop still further the same subject, by considering a few additional examples; and first I will ask your attention to those evidences of design which are to be found in the relations of the atmosphere to light and heat. Here, however, I am met by a difficulty. In order to explain fully these relations it would be necessary to develop from first principles the sciences of optics and thermotics, and to do this in a popular manner would require several lectures. These sciences furnish some of the most wonderful evidences of design which are to be found in nature, and I have no doubt will be given their appropriate place in this series of lectures. Without, therefore, attempting any detailed explanations, I will merely bring before you a few facts, drawn from these departments of knowledge, which illustrate the adaptations of the density of the atmosphere to its appointed ends.

Transparency of the atmosphere.

The atmosphere, although very much more pervious to light than any kind of solid or liquid matter, is far from being perfectly transparent. Indeed, the reverse is sufficiently evident from our daily experience. Every one has noticed that distant objects appear less distinct in proportion as they are removed, their colors become fainter, the contrast between light and shade less marked, and that they seem as if covered with a pale blue veil. This effect, always noticed on distant mountains, is owing to a partial absorption of the light while passing through the atmosphere; for, were the passage of the rays wholly unimpeded, all objects, although reduced in size in proportion to their distance, would appear equally distinct, and their colors equally brilliant.

The transparency of the atmosphere differs very greatly under different circumstances, but it has been estimated, that under the most favorable conditions, at least thirty per cent of all the light coming from the heavenly bodies is absorbed before reaching the surface of the earth, and in our latitude at this season, even when the sun is on the meridian and the sky clear, fully one half of his rays are thus spent. Do not suppose, however, that all the light so expended is lost. Quite the contrary, for every particle of the atmosphere, illuminated by the sunbeam, becomes itself a new centre of emission, radiating the light in every direction.

Diffusion of the sun's rays.

This diffusion of the sun's rays is the cause of that wonderful effect which we term daylight. I say wonderful effect, for although so familiar, it is one of the most remarkable results of skilful adaptation and infinite wisdom. The very daylight which streams in at the open windows of our houses, filling them with cheerfulness, and penetrating to their inmost recesses, which enlivens the whole landscape, and which bars and bolts cannot wholly exclude even from the prisoner's dungeon, is another evidence of the adjustment of the density of the atmosphere to the constitution of man.

In order that you may see the force of this evidence, let us inquire what would be the effect of the sun's light if there were no atmosphere, assuming, for the sake of illustration, that man could exist under such conditions. The known principles of optics will furnish an answer to our question. They inform us that while all objects directly illuminated by the sun would shine so brilliantly as to dazzle the eye, everything else would be in complete darkness. The interior of our dwellings would be at midday as dark as during the darkest night; and the contrast between light and shadow would be stronger than that between the most brilliant sunlight and the blackest darkness now known. Our eyes, as little fitted to such conditions as our lungs, would be blinded by the sudden alternations, and distinct vision would be rendered impossible.

Thus it appears that the atmosphere is as much an essential condition of our seeing as of our breathing. But more than this, the immeasurable pleasure which we derive from our sense of vision depends upon the exact adaptation of the density of the atmosphere to the organization of the eye. Similar effects to those just noticed, although of course on a greatly diminished scale, would result, were the density of the atmosphere

Effect on the landscape.

materially diminished. This is not a matter of theory, for such effects have been frequently noticed on the summits of high mountains, where the air is much rarer than at the sea level. On the top of Mont Blanc the sky has a blackish hue, and the stars are seen at midday. The glare of the direct light is unsupportable to the eye, and even the reflection from the snow blisters the unprotected skin, while at the same time the contrast between light and shade, unnaturally increased, gives to all near objects a peculiar and ghastly aspect. This effect results, it is true, from a very great diminution in the density of the air; but when you reflect upon what delicate contrasts of light and shade the beauty of a landscape depends, — the clearness of the foreground, the gray of the middle distance, and the tender purple of the distant hills all blending into one harmonious whole, — you can appreciate how slight a change would disturb the result, and deprive the sense of beauty of its purest enjoyment.

I have thus far spoken only of the influence of the atmosphere in softening the intensity of the rays of light, and in diffusing their action; but the atmosphere has also, under certain conditions, the power of decomposing the sun's rays, and thus producing, not only those displays of gorgeous tints which we witness in the sunset clouds, but also the pure blue which colors the dome of heaven.

In regard to the precise means which are employed by nature to produce these results, scientific men are not agreed. It has been proved that the blue color of the sky is seen by reflected light, and it is probable that the color is caused by repeated reflections of the sun's rays from the surfaces of the innumerable small water-bubbles which are constantly floating in the atmosphere. You have all noticed the blue color of the soap-bubble shortly before it breaks. This color is caused by the action of the very thin film of water in decomposing the light reflected from its surface, and it is supposed to be an action of the same sort, only very much increased by repeated reflections, which gives to the sky its azure hue.

Blue color of the sky.

While the blue color of the sky appears to result from changes in the white light of the sun caused by reflection, it is equally probable that the sunset tints arise from changes in the same white light caused by an unequal absorption of its different colored rays during their transmission through the atmosphere. Here, again, the vapor in the air is supposed to be the active agent; and the theory is, that the tints are produced while the vapor is condensing into clouds,— a change which naturally occurs at sunset. But this is a mere theory, and our whole knowledge on these subjects is very imperfect.

Sunset tints.

So far, however, as our present argument is concerned, it is not essential that we should understand exactly how these glorious results are obtained. It is sufficient that we are constantly enjoying their beauty, and that we know they are owing to the peculiar constitution of the atmosphere. When future discoveries shall bring to light the methods, at present secret, by which nature gilds the sunset clouds and covers our beautiful dwelling-place with its canopy of blue, we shall unquestionably find fresh evidences of God's wisdom; but even now, when ignorant, perhaps, of these hidden causes, we have that which is far more excellent, the most conclusive evidence of His goodness and love. Our Father has not only adapted the atmosphere to the wants of our bodies, and made it conducive to our physical enjoyment, but he has also made it the scene of the highest beauty,—a beauty which fills the longings of our souls and calls forth their noblest and purest aspirations. Man, sinful as he is, cannot look up into the pure blue of heaven without a sense of reproach, and the feeling that it is a fit emblem of the kingdom of purity and peace. And when the setting sun lights up the evening altar in the west, who can repress the rising prayer of devotion, and hesitate to believe, with the child, that his Heavenly Father is smiling behind the clouds? There is a depth to the beauty of nature which man cannot fathom. Poetry cannot describe it, and the highest art only displays its weakness when it attempts to copy it. The savage feels that it is immeasurably above him, and worships it. The artist seeks to attain it, but the more he strives, the more it surpasses his power, and he dies disappointed, unless, happily, he finds that the perfect ideal has been realized only in Christ, and thus through nature is led up to nature's God. Yes! the beauty of nature is in the Infinite Presence it conceals, and, unconsciously though it may be, it is the spirit, not the matter, which the artist loves.

Such are some of the evidences of design which we discover in the relations of the atmosphere to light. Let us now examine some of its relations to heat, which we shall find not less instructive. It was formerly supposed that the rays of heat, although accompanying the luminous rays in the sunbeam, were essentially different from those of light. But it is now almost universally believed that the rays of heat differ from those of light only, at most, as one color differs from another, and that even the same rays, which, falling on the retina of the eye, excite the sensation of light, when falling on the nerves of feeling may excite heat. But what, you may ask, is the difference between the different colors? The subject is somewhat abstruse, but if you will follow me attentively for a few minutes I will try to make it intelligible.

Relations of the atmosphere to heat.

Every one who has dropped a stone into the water of a still lake has noticed the system of waves which, with its ever-increasing circles, spreads in every direction from the stone; but all may not know that when

Wave motion, illustrated by waves of sound.

two stones are struck together in the air a similar system of aerial waves spreads, in ever-widening spheres, through the atmosphere, and that it is these waves breaking on the tympanum of our ears, like the waves of water on a sand-beach, which produce the sensation we call sound. Two stones thus struck together give rise to waves of unequal size, following one another at irregular intervals; and such waves produce an unpleasant sensation on our auditory nerves, which we call noise. But if, instead of striking together two stones, we set in vibration the string of a piano-forte or the reed of an organ-pipe, we excite a system of waves, all of equal size, and succeeding one another with perfect regularity, and these breaking on the ear produce by their regular beats what we call a musical note. If the waves follow one another with such rapidity that one hundred and twenty-eight break on the tympanum every second, the note has a fixed pitch, called in music C natural. If the waves come faster than this, the pitch is higher, and if less rapidly the pitch is lower. What you are all familiar with as the pitch of a musical note depends, then, on the rapidity with which the waves of sound strike the ear, and may evidently be measured by the number of waves breaking on the tympanum in a second.

Our ears are so constituted that they can hear a musical note only when within certain fixed limits of pitch, differing to a slight extent with different individuals. The deepest bass note, which can be heard, as such, by a good ear, is produced by about eight waves in a second. If the waves strike less rapidly than this, they are perceived as distinct beats, and beginning at this note the musical scale ascends to a note caused by twenty-four thousand waves a second, which is the highest note perceptible by human sense. The range of a piano generally extends from a note produced by sixteen waves in a second, to one caused by one thousand and twenty-four waves in a second, as is shown by the accompanying table.

Notes.	Length of waves in feet.	Number of waves striking the ear in one second.
C — 3	70	16
C — 2	35	32
C — 1	17.5	64
C_1	8.75	128
C_2	4.375	256
C_3	2.178	512
C_4	1.093	1024

Name of Note	C_1	D_1	E_1	F_1	G_1	A_1	B_1	C_2
Number of Waves	128	144	160	$170\frac{2}{3}$	192	$213\frac{1}{3}$	240	256
Ratio of each number to that of Note C_1	1	$\frac{9}{8}$	$\frac{5}{4}$	$\frac{4}{3}$	$\frac{3}{2}$	$\frac{5}{3}$	$\frac{15}{8}$	2

Sounds of the highest pitch, like the cry of some insects, become disagreeable, and by some persons cannot even be distinguished. It is quite possible to produce a sound, which, though painfully shrill to one person, shall be entirely unheard by another. Professor Tyndall, in his very interesting work on the glaciers of the Alps, relates an instructive anecdote of this sort, which I give in his own language.

"I once crossed a Swiss mountain in company with a friend; a donkey was in advance of us, and the dull tramp of the animal was plainly heard by my companion; but to me this sound was almost masked by the shrill chirruping of innumerable insects, which thronged the adjacent grass; my friend heard nothing of this, it lay quite beyond his range of hearing."

There may, therefore, be innumerable sounds in nature to which our ears are perfectly deaf, although they are the sweetest melody to more refined senses. Nay, more, the very air around us may be resounding with the hallelujahs of the heavenly host, when our dull ears hear nothing but the feeble accents of our broken prayers.

We have been studying, my friends, the nature of sound, in order to comprehend more readily the nature of light and heat, for the phenomena included under these names are produced, like the phenomena of sound, by waves, not, however, by waves in the air, but by waves in a medium which is as much more subtile than air as air is more subtile than water, — indeed, a medium so exceedingly thin that it eludes all our powers of chemical analysis; but which, as we assume, pervades all space, and this, too, whether the space is filled or not, at the same time, by other forms of matter. We call this medium ether, and through it the waves of light speed

Waves of light.

with an inconceivable rapidity. Sound travels 1,100 feet in a second, but a wave of light spans 192,000 miles in the same time, and starting from the sun on its journey of unnumbered years, to Sirius or Arcturus, leaves the whole solar system behind in a single hour.

Yet great as is the difference of velocity, the analogy between sound on the one side and light or heat on the other is complete. Every luminiferous body, like this candle-flame, excites in the tenuous ether a system of waves, which spread in ever enlarging spheres with the immense velocity just described; and it is these little billows which, passing through the humors of the eye, and breaking on the retina, produce the sensation we call light, or, falling on the skin, excite the less delicate nerves of feeling, and cause the sensation of heat.

Moreover, the difference between colors is of precisely the same kind as the difference between notes. Red, yellow, green, blue, violet, &c., are names we give to sensations caused by waves of ether breaking at regular intervals on the retina. Color corresponds to pitch, and — what may seem to you incredible — we are able to calculate from actual measurements the number of waves of ether which must break on the retina in a second in order to produce the sensation of a given color. Here are some of the numbers, and, extravagant as they appear, they are the sober results of science, and have been as accurately determined as the magnitudes and distances of astronomy.

Cause of color.

Colors.	Length of waves * in parts of an inch.	Number of waves in a second.
Red . . .	256 ten-millionths . . .	477 million million.
Orange	240 "	506 " "
Yellow	227 "	535 " "
Green	211 "	577 " "
Blue	196 "	622 " "
Indigo	185 "	658 " "
Violet	174 "	699 " "

It is actually true, that when we are receiving the sensation of red there are no less than 477 million millions of ether waves breaking on the retina of our eyes every second. And more than this, we have measured the length of these waves, and we know that the length of a wave of red light from crest to crest is $\frac{1}{39000}$ of an inch. By examining the table you will also discover that the sensation of red, as compared with other colors, results from the smallest number of waves, and that these waves are comparatively large. On the other hand, the sensation of violet is caused by the largest number of waves, which, however, are proportionally small in size. The red light, therefore, corresponds to low, and the violet to high notes of

* A given wave-length corresponds to each point on the line of the solar spectrum to be described further on. The numbers given in the table are to be regarded merely as the mean values for each color, measured at points on the spectrum, marked by certain prominent dark lines, called Frauenhofer's lines. The solar spectrum, when formed by a beam of light admitted into a dark chamber through a narrow slit, is seen crossed by thousands of these lines, which have a fixed position, and therefore serve to mark definite points on this otherwise continuous band of blending colors.

music, and between these extremes there exists every gradation of pitch, which is here manifested in color.

Waves of all the dimensions given in the table, together with waves of every possible length between certain extremes, — which are far wider than those indicated above, — move together in the sunbeam, and their combined impression produces the sensation of white light. We have a very simple way of analyzing the sunbeam and separating its different color-producing waves. The method consists in passing the sunbeam through a glass prism. The prism has the power of bending the beam from its rectilinear direction; but it does not change the direction of the motion of all the waves to the same extent. The longer waves, which give the sensation of red, are bent from their course much less than the shorter waves, which produce the sensation of violet, while waves of an intermediate length take a course between the two. Hence, after emerging from the prism the directions of the different waves diverge, and if we receive the beam of light thus analyzed, on a screen, the various color-producing waves strike the screen at different points of a continuous line. A more or less narrow band on the screen will thus be illuminated with lights of different colors in the order, Red, Orange, Yellow, Green, Blue, Indigo, Violet, and this beautiful phenomenon is familiar to almost every one under the name of the solar spectrum.

The solar spectrum.

Analogy between colors and musical notes.

Here, where we have the whole scale of colors spread out before us, the analogy of light to sound becomes still more evident. As there are persons who cannot hear the shrill sound of some insects, so there are many who cannot see certain colors of the spectrum, and as there are unquestionably innumerable sounds in nature which are inaudible to our ears, so there are innumerable waves in the ether which are powerless to produce the sensation of light. Moreover, singular as it may seem, we have more palpable evidence of the existence of these non-luminiferous waves than we can obtain in the case of sound. There are waves in the ether far smaller, and undulating far more rapidly, than those which produce violet light; so small that they do not even jar the nerves of the retina, but which, nevertheless, breaking on the prepared plate of the photographer, leave there an impression which, developed by his skill, becomes a beautiful copy of nature or of art. On the other hand, there are waves in this same ether so large that the delicate retina cannot vibrate in unison with their rough beats, but which, nevertheless, breaking on the surface of the skin, disturb the coarser nerves of feeling, and produce the glow of heat. Most of the waves which impress the optic nerve will also affect the nerves of feeling; but the reverse is not true, for many of the waves which produce the sensation of heat are far too large, and undulate too slowly, to set in vibration the retina of the eye.

I hope that I have been able to make clear two points, — first, that light and heat are forms of motion; second, that the differences in the phenomena which have been referred to these two agents are simply different sensations or different effects* produced by the same wave-motion. It would be highly interesting in this connection to examine the wonderfully delicate adjustments and to follow out the peculiarly intricate motions which concur to produce the phenomena of light and heat; for they are in themselves most striking illustrations of the wisdom of the Creator. But this would lead us too far from our proposed plan, and I must content myself with the few facts already given, which were necessary to illustrate the relations of the atmosphere to the thermal conditions of our globe.

From the principles stated, it is evident that the atmosphere must act in diffusing heat just as we have seen that it acts in diffusing light.

* The effects of expansion, melting, evaporation, the permanent elasticity of gases and vapors, and many other phenomena, formerly referred to the action of a peculiar agent called heat, are now supposed to be the result of the motion which the ether-waves communicate to the material particles of the bodies on which they strike or through which they are transmitted. To understand this, we must remember that the molecules, even of the densest solids, are supposed to be separated from each other by comparatively large spaces filled with ether, through which the waves of heat and light may move more or less freely, just as the waves of air pass between the branches in a forest. Moreover, as the waves of air impart motion to the branches of the trees, and afterwards are kept in motion by the waving boughs, so also the material particles of a body may set in motion the waves of ether, or receive motion from them in return.

Indeed, this effect is one of the thousand conditions on which the existence of organic life depends. Were it not for the influence of the atmosphere, the greatest extremes of temperature would be produced by the alternation of day and night, and even were the density of the atmosphere reduced only one half, the variation would be so great as to render the existence of the higher forms of organic life impossible, except, perhaps, in the more favored regions of the earth.

The atmosphere diffuses the sun's heat.

But not only does the atmosphere diffuse the heat of the sun's rays, it also acts, and even more effectually, in retaining on the surface the heat which the earth is constantly receiving from that great central luminary. The atmosphere has been compared to a mantle, and the comparison is just; for, like a huge cloak, it envelops the earth in its folds, and protects it from the chill of the celestial spaces through which we are rushing with such frightful velocity. In order to understand how a thin and transparent medium like air can thus act to keep the earth warm, we must recur to some of the facts established above.

The atmosphere retains the sun's heat.

As the ether-waves, breaking on the eye more or less rapidly, produce the different sensations of color, so when breaking on the skin they occasion analogous differences in the sensation of heat, which, although not so accurately distinguished, because the sense is less delicate, nevertheless are as real as the difference between

Heat of low and high intensity.

a low and sweet musical note, and one that is high and shrill. There are waves of heat which break upon our nerves of feeling like the shrill cry of the cricket on the ear, and seem to penetrate to the very brain, while there are others which fall like the low tones of an organ, diffusing throughout the system a genial glow. Such, for example, is the difference between the heat from a hard-coal fire and that from a steam radiator. The waves of the first sort, from their small size and rapid motion, can readily pass through glass and other transparent media, when the large waves with their slow motion are in a great measure stopped.

Now it is found that the sunbeam is chiefly made up of waves of the first class, which are therefore able readily to penetrate the atmosphere and warm the surface of the earth. The earth thus warmed becomes itself a hot body, surrounded by an intensely cold space, and, like any other hot body, tends to lose its heat by radiation. But the waves of heat which the earth* sets in motion are of the second class, long and slow undulations, and these are in great measure arrested by the atmosphere; indeed, as experiments have proved, they are chiefly absorbed by the lower strata,† in which we live and move.

* The pitch, if we may so speak, and penetrating power of the heat-waves depend on the temperature of the body by which they are set in motion, and in proportion as the temperature rises the pitch is higher, and the penetrating power greater.

† Professor Tyndall has shown that this effect is due almost entirely

Thus it is that the atmosphere keeps us warm; and if you desire further proof of the correctness of these experimental deductions, ascend any high mountain, and, as the thickness of the aerial covering above you is diminished by the elevation, you will find that the chill increases, vegetation slowly disappears, and before long you will reach a region of eternal snow and ice. It is true that there are other causes acting to lower the temperature at high elevations, but the one just noticed is by far the most important, as well as the primary cause. The effect of the atmosphere is precisely similar to that of the glass sashes of a hothouse. The glass, like the atmosphere, allows the rapidly undulating waves of the sun to pass,* but almost entirely arrests the large and slowly undulating billows, which radiate from the vegetation within. They are each, in fact, *a trap to catch the sunbeam.*

The atmosphere distributes the sun's heat.

The atmosphere not only thus acts in diffusing the sun's rays, and retaining the heat which they bring to us, but it also subserves an equally important end in distributing their genial warmth over the whole surface of the earth, thus moderating the climate

to the aqueous vapor in the atmosphere, which is present in greatest quantity in the strata nearest to the earth.

* In the sunbeam, as it passes through space, there are undoubtedly waves of low pitch in abundance, but these are almost entirely arrested by the atmosphere before reaching the surface of the earth. It has been estimated that of the heat the earth receives from the sun about one third is thus absorbed.

of the temperate zone, and mitigating the intense heat of the tropics. Air, like all gases, is expanded by heat, and thus rendered specifically lighter, and on this simple principle all our methods of warming and ventilating are based. When now it is remembered that the atmosphere under the tropics must become more intensely heated by the vertical rays of the sun than it is in the temperate zones, the result will be obvious. The heated air rises, and the cold air rushes in from the north and south to take its place. Thus, two general currents are excited in the aerial ocean of each hemisphere, one on the surface of the earth tending towards the equator, and another, higher in the atmosphere, tending towards the poles. These currents, however, do not blow due north or south; for many causes combine to turn them from their primitive directions.

Trade winds.

In the first place, the rotation of the globe on its axis imparts to objects on the surface a motion from west to east, varying in velocity, from nothing, at the poles, to the speed of a cannon-ball, at the equator. In consequence of this, a mass of air moving towards the equator is constantly arriving at a point on the surface of the earth, which is moving towards the east more rapidly than the point it has just left; and as, in virtue of the law of inertia, the moving mass cannot accommodate itself instantaneously to the increased velocity, it is left a little behind, — that is, a little to the west, at every step. Hence, the lower or polar currents bend more and more towards the west as they approach the equator, acquiring in the northern hemisphere a south-westerly, and in the southern hemisphere a north-westerly direction; and the currents of the two hemispheres, meeting at the equator, combine to produce the great trade-wind, which, in the Pacific Ocean, blows constantly from the east to the west, and would blow regularly in this direction all round the globe if the continents did not intervene to disturb its course at various points.

The effect of the earth's rotation on the current of warm air which flows from the equator in the upper atmosphere, must evidently be the reverse of that just described, bending it constantly to the east, and giving to it in the northern hemisphere a northwesterly, and in the southern hemisphere a southeasterly direction. But the upper and lower currents do not long retain this relative position; for, as the first comes northward, it gradually sinks, and, long before reaching this latitude, touches the surface of the earth. Then, of course, it comes in collision with the current from the north, and here a strife for the mastery ensues. Sometimes the one and sometimes the other prevails, and this alternating ascendency is one of the chief causes which render the winds of temperate climates so irregular.

Again, the unequal heating effect of the sun's rays on the earth, as compared with the sea, combined with the irregular distribution of land and water over the surface of the

Causes of irregularity.

globe, tends to complicate still further the motion of the aerial currents. For reasons which will hereafter appear, the land is more heated by the sun's rays than the sea, when under the same conditions, and, on the other hand, as soon as the sun is withdrawn, it cools more rapidly. Hence, on an island in a warm climate we generally have, during the day-time, a current of heated air rising from the surface of the earth, and a current of cooler air flowing in on all sides from the ocean to take its place, while after sunset the land soon cools below the temperature of the surrounding ocean, and the current is reversed. Thus is produced the daily alternation of land and sea breezes, which is so familiar to every one who has visited the tropics, where these phenomena are most strongly marked.

Monsoons.

Quite a similar reciprocal action between the continents and the great ocean is caused by the alternation of seasons, and of this the monsoons of the Indian Ocean are a remarkable illustration. This *mediterranean* ocean, shut off from the influence of the general trade-winds by the great continental masses which surround it, has a system of aerial currents peculiar to itself, blowing six months of the year in one direction and six months in the other. These are set in motion by the unequal heating of the continents of Asia and Africa during the extreme seasons. In the months of December, January, and February, the part of Africa south of the equator is exposed to the vertical rays of a summer's sun, while the countries of Southern Asia are feeling the comparative cold of their winter. The natural consequence is, that a stream of cold air rushes across the Indian Ocean to feed the intensely-heated current which is rising over the burning plains of Africa, producing a strong northeasterly breeze, which is the winter monsoon of India. When, however, the sun comes north of the equator, all these conditions are reversed. The ocean air now rushes to the more heated plains of India, and the summer monsoon sets in, which blows from the southwest, the change from one to the other being always attended by variable winds and furious storms.

Lastly, the position of mountain chains and the configuration of the continents, which break and turn the winds, or open to them a freer channel, have an important influence in determining the direction of the aerial currents on the earth. But we have not time for further details; they are given in all works on physical geography,* and the student of natural theology will find that subject rich in illustrations of God's wisdom and power. We have already become sufficiently acquainted with the general plan to understand how the atmosphere acts in equalizing the climate of the globe. The aerial currents which come to us from the south bring with them the heat of the tropics, and distribute it over the temperate zone. As they blow

* See "Earth and Man," by Professor Arnold Guyot.

from the southwest, they naturally exert the greatest heating power on the western coasts of the continents, and this is one great cause of the well-known fact that the climate of Western Europe is so much milder than our own, and the climate of California and Oregon so much warmer than that of the corresponding latitudes on the eastern coast of Asia. Moreover, the sea-breezes on islands and along sea-coasts, the monsoons of the Indian Ocean, and other local currents, all combine, as our theory shows, to produce the same general result, cooling such regions of the earth as from any cause have become overheated, and transferring the warmth to places where it is more needed. Just as the heat of burning fuel is diffused over a whole building from the furnace by the currents of air it sets in motion, so the sun's heat is diffused over the earth from the tropics by the great terrestrial currents we have so briefly described. Indeed, as already stated, in all our methods of heating, we merely apply, on a small scale, the same general principles which are at work around us in the atmosphere.

Capacity of the air for heat.

But, although the heat of the sun might set in motion these aerial currents, they would have but little effect in warming our northern climate, were it not that the air has been endowed with a certain capacity of holding heat. All substances possess this capacity to a greater or less degree, but the differences between them are very large. Thus the amount of heat required to warm a pound of water is ten times greater than would be required to raise the temperature of a pound of iron, and thirty times greater than would be required to raise the temperature of a pound of mercury to an equal extent. Hence, under the same conditions, a pound of water may be said to contain ten times as much heat as a pound of iron, and thirty times as much as a pound of mercury; or, again, in other words, the capacity of water for heat is ten times greater than that of iron, and thirty times greater than that of mercury. The capacity of air for heat is, weight for weight, about twice as great as that of iron, and although only one fifth as great as the capacity of water, it is yet greater than that of most other substances. The point, however, to which I wish to direct your attention, is the fact that this capacity is exactly adjusted to the office which the air has been appointed to fill. Were the capacity of the air less, the hot air from the tropics would bring to us proportionally less heat; were it greater, the reverse would be the case; and in either event, the distribution of temperature on the earth would be changed. To what extent such a change would affect the general welfare of man, it is impossible to determine; but when we consider how far the history of man has been influenced by climate, it will appear that the present distribution of the human race — the existence, for example, of a large and influential city in this place, — may be said to depend on the adjustment of the capacity of the atmosphere for heat; and yet it depends no less on ten thousand other conditions, many of them far more important than this. How truly, then, it may be said, that even here on earth we live in "a city which hath foundations, whose builder and maker is God"!

Relations of the atmosphere to electricity.

Such are a few of the more obvious marks of design, which may be discovered by studying the relations of the atmosphere to light and heat. I might here close one division of my subject; but I should fail to give you an adequate idea of the wonderful play of physical forces in the atmosphere, were I to leave out of view that mighty agent which charges the artillery of heaven and feeds the flaming torches in the northern sky. It is true that the atmospheric relations of electricity are very imperfectly understood, and the important ends which it undoubtedly subserves in the economy of nature almost entirely unknown. We cannot, therefore, expect that they will furnish us with many additional illustrations of the Divine attributes; but since electrical phenomena play so conspicuous a part in the atmosphere, and must have been included in its plan, they should not certainly be overlooked if we would gain a general idea, however imperfect, of the whole design.

Nature of electricity.

Of all the assumed agents of nature there is hardly one which is so little understood, and yet has been so carefully studied, as electricity. To the uneducated it affords the convenient explanation of most obscure phenomena, while with men of science it is the object of much laborious investigation and careful theorizing. The study of its phenomena has been fruitful in the discovery of facts; but it has as yet led to but few general principles, and has furnished only a meagre explanation of those grand displays of nature in which it seems to be such an important agent.

In regard to the nature of electricity, we are entirely ignorant. The phenomena of light and heat * admit, to say the least, of an intelligible explanation, and can be referred to a dynamical origin; but in the case of electricity we are obliged to be content with collecting facts, and must await the further progress of science to reveal the now hidden cause. I am well aware that electricity has been regarded as a very *rare and subtile fluid*, and that this theory has not only afforded a plausible explanation of most of the phenomena of statical electricity, but also that the numerical results based upon it have been most remarkably verified by experiment. Yet nevertheless, although the theory may still be used as a convenient frame in which to exhibit the facts, there are but few investigators of the present day who would claim for it more than a very partial foundation in truth, and most would reject it altogether as utterly untenable.

The fundamental facts of electricity were known to the ancients, and are familiar to every one. If a stick of sealing-wax or a glass tube is rubbed

* See a recent work by Tyndall on Heat as a Form of Motion.

with a warm silk handkerchief, it becomes, as we say, electrified, and in this condition has the power of attracting pieces of paper or any light particles of matter. When the scientific men of the last century came to examine these phenomena more carefully, they found that the handkerchief was also electrified and thrown into a state differing from that of the glass in the one case and that of the resin in the other, very much as the north pole of a magnet differs from its south pole. They found also that the resin was electrified oppositely to the glass, and they hence concluded that there were two kinds of electricity, which they distinguished by the names resinous and vitreous, or positive and negative. They also discovered that this agent could readily be drawn off from electrified bodies by the metals, but only with difficulty, if at all, by such materials as India-rubber, glass, resin, or silk, and they were hence led to divide substances into conductors and non-conductors of electricity. A good conductor, when insulated by non-conductors, was found to retain for a short time the electricity it had received from the electrified glass or resin, although the charge was soon dissipated by the surrounding air, especially when moist. By bringing in the aid of machinery, and thus increasing the surfaces of friction, it was found possible to enhance very greatly the effects obtained with a glass tube; and this was the origin of the electrical machine. This familiar instrument is merely a mechanical contrivance for rubbing together glass and

Fundamental facts.

silk, with two insulated metallic conductors for receiving the two kinds of electricity thus generated. If the hand or a metallic knob was brought near the prime conductor of the machine when highly electrified, it was found that a luminous discharge followed, which was termed an electrical spark; and it was found possible by means of a glass jar, coated inside and outside with some metallic leaf, called a Leyden jar, to accumulate the two electricities in such large quantities that, when allowed to flow together, the discharge was capable of producing violent mechanical action, similar to that of lightning, although on a vastly reduced scale. It was also discovered that electricity passes readily through the greatly rarefied atmosphere in the receiver of an air-pump, causing a luminous effect similar to the aurora borealis. Lastly, it was observed that electricity readily escapes into the atmosphere from a pointed conductor, and, conversely, that a heavy charge can be silently and harmlessly drawn from an electrified body by holding near it the point of a needle. By attaching a pointed conductor to a boy's kite, Franklin succeeded in drawing an electric spark from a thunder-cloud, and having thus established the identity between atmospheric and friction electricity, he erected the pointed rod, which protects our dwellings against the lightning's stroke.

More recently it has been discovered that friction is by no means the only source of electricity, and it seems probable that no change, either chemical or physical, takes place in Nature without some manifestation of this agent. It was at first supposed that there were several kinds of electricity, which were named thermo-electricity, magneto-electricity, voltaic electricity, and animal electricity, according to the nature of the process in which the electrical action was developed; but it is now universally conceded that all are only different manifestations of the same agent, and most investigators believe that electricity will in time be shown to be a form of molecular motion analogous to that which produces the phenomena of light and heat, although it has not as yet been found possible to frame a comprehensive and intelligible theory based upon this hypothesis. Again, it has been found that friction is a far more general source of electricity than was at first believed. In fact, electrical phenomena appear to be a constant result of friction, whatever may be the nature of the substances rubbed. Thus it is developed by blowing air over glass, and the hydro-electric machine, one of the most effective means of generating electricity we possess, owes its surprising energy to the friction of globules of water against the sides of the vent-cock of a steam-boiler.*

When, now, we consider that the air is always rubbing over the surface of the earth, at times with

* This machine consists simply of a small steam-boiler insulated on glass pillars, having a peculiarly-constructed vent-cock and provided with suitable metallic conductors for receiving the electricity. The steam, as it escapes under high pressure, becomes filled with globules of water which rub against the sides of the vent-tube, and this is so shaped as to facilitate their formation.

great rapidity, we shall not be surprised to learn that both bodies are constantly in an electrified condition, the earth being generally charged negatively, and the atmosphere positively. Even in fair weather it is always possible to detect the presence of free electricity in the atmosphere; and during a storm, when clouds filled with vesicular drops of water are hurried over the surface, grinding against the hills and the trees, or against each other, the atmosphere becomes a vast hydro-electric machine, whose sparks are the lightning, and the noise of its discharges the thunder. Again, the various chemical and physical changes which are going on around us, — such as the vital processes of animals and plants, the combustion of fuel, volcanic action, the evaporation of water, — all undoubtedly add to the electrical excitement of the atmosphere, and more or less modify the result. It is not important for us, however, to study the action of each one of these causes; for we have, probably, in the friction of moist air driven by the winds, the chief source of atmospheric electricity; and when we consider the amount of friction which must attend the rapid motion of storm-clouds, or of a tornado through the atmosphere, the wonder is, not that an occasional thunderbolt should kindle a conflagration, or even cause a death, but that every storm does not lay waste the earth along its fiery track. Moreover, when we appreciate the vastness of the scale on which the electrical machine of nature is constructed, the thunder-storm ceases to surprise

Atmospheric electricity.

us, and only calls our attention to those beneficent provisions by which we and our race are saved by a constant miracle from the fate of the cities of the plain. That the atmospheric electricity was designed to subserve many important and beneficent ends, the whole analogy of nature compels us to believe; but while our present ignorance conceals them from our sight, we may still discover evidence of God's goodness and wisdom in those simple provisions by which the atmosphere is preserved from violent or frequent electrical excitement, and its charge drawn down harmlessly to the earth.

Protection against lightning.

Since the atmosphere is at best a very poor conductor, the electricity developed by the processes just considered tends to accumulate, and under peculiar conditions the clouds may become so highly charged, that at length the pent-up power acquires sufficient force to break through all barriers, and the lightning dashes to the earth, crashing, rending, and burning on its way. To guard his roof from its destructive action, man erects the lightning-rod, whose bristling points quietly drain the clouds, or, failing to do this, receive the charge, and bear it harmlessly to the earth. But ages before Franklin pointed the first rod to the storms, the merciful Parent of mankind had surrounded the dwellings of his children with a protection far more effectual than this; for, since the creation of organic life, every pointed leaf, every twig, and every blade of grass have been silently disarming the clouds of their destructive weapon. It is difficult to improve upon nature, and man constantly finds that in his best inventions he has been anticipated from eternity by a greater Inventor than he. So not long after Franklin had discovered the efficacy of metallic points in dissipating a charge of electricity, and had applied the principle in constructing the lightning-rod, it was found that a common blade of grass pointed with nature's exquisite workmanship was three times as effectual as the finest cambric needle, and a single twig far more efficient than the metallic points of the best-constructed rod. When, now, you reflect how many thousand of these vegetable points every large tree directs to the sky, and consider what must be the agency of a single forest with its innumerable twigs, or of a single meadow with its countless blades of grass, and then when you remember that these are only subsidiary to those vast lightning-conductors, the mountain-chains, whose craggy summits pierce the clouds themselves, and still further, when you learn that the rain-drops and snow-flakes have been made good conductors, so that during the storms a bridge for the lightning is thrown across from the clouds to the earth, you will see how abundant the protection is, and with what care Providence has guarded us from the destructive agent. It is only under unusual circumstances, when electricity is developed more rapidly than it can be dissipated through these numberless channels, that a violent discharge takes place, and if then it tears, burns, or kills, it also reveals the Merciful Hand which constantly spares. Moreover, through this servant of his pleasure, God is constantly educating and elevating his creatures. In the wild corruscations of the lightning, and in the reverberating roll of the thunder, Nature exhibits one of her grandest aspects, before which man is awed and terrified, while he admires and adores; and also when, through the cold, dry air of the arctic region the electric charges shoot down to the earth in tremulous flashes, then Nature lights up her grand displays of northern fires, which enliven the long night of the arctic winter, or in this more favored climate excite the curiosity and call forth the admiration of all.*

Distinctive marks of the design of nature.

I must here conclude this very imperfect sketch of the physical adaptations of the atmosphere to the ends it subserves on the earth. We studied in the first place its aeriform condition, and found that its density not only forms an essential part of the scheme of organic nature, but also is closely related to the dimensions of the solar system. In this Lecture we have studied the relations of the atmosphere to light, heat, and electricity; and although we have been able only to glance at some of the more prominent features in

* I am indebted for many of the above illustrations to an admirable paper on atmospheric electricity in the American Almanac for 1854, by my friend and colleague, Prof. Joseph Lovering.

these wonderful displays of creative energy, we have found, wherever we turned, abundant illustrations of the wisdom, power, and goodness of our God. I trust that you have been impressed by the vastness, the complexity, and yet the simplicity and harmony of the whole design, for these are the chief points which I have endeavored to set forth. But O how imperfect any conception which I can give you must be! This atmosphere is sustained in the proper working of all its parts only by the exact balancing of a thousand conditions. Attempt to make yourself acquainted with these conditions, and, disregarding those which you recognize at once as surpassing human intelligence, study only such as are thoroughly understood and universally admitted to have been primary conditions in the plan of nature before the atmosphere could exist as it is. This is not an impossible task. It would require years of study, and it would lead you into every department of physical science, but the result would well repay your labor. You would find it easy to follow out any one line of the conditions, until it became lost in the obscurity of the unknown; but to form a conception of the simultaneous working of all in their varied relations, or even of two or three, you would soon discover to be a hopeless task. The complication of this wonderful machinery so far transcends man's mental power, that to understand its combined action is simply impossible. But although you would thus be made most keenly sensible of the limits of your human powers, you would also be filled with gratitude for the great privilege you enjoyed of studying this divine mechanism, even if you could understand its workings only obscurely and in part.

Paley has compared the mechanism of nature to a watch, and, so far as the argument for design is concerned, the analogy is perfect. We must never forget, however, that there is an essential difference between the scheme of nature and the most complicated human mechanism. I have seen a carpet-loom weaving a pattern composed of twelve different colors, and, as I watched the shuttles of various colored yarns which were selected by the hands of the machine with unerring certainty, and thrown through the warp, it seemed as if the very iron were endowed with intelligence, and the impression was one of wonder and bewilderment. To comprehend such complexity appeared impossible; but the more I studied the details of the machine, the more thoroughly I understood the mode of its action, until at last the wonder vanished; and although not ceasing to admire the skill of the inventor, I felt I had comprehended the whole, and could even conceive of the mental process by which such a wonderful combination of means had been made and adjusted to produce the desired end. The artist was ingenious, but the machine was still human.

How different it is with the mechanism of nature! Here, also, it is true, the more we study, the more we understand the workmanship; but then we never reach the limit. The more our powers of thought and observation are developed, and the more our experience is enlarged, the more the field of possible knowledge expands before us. The larger our attainments, the less we seem to know.

We still recognize the unmistakable marks of intelligence in the design, but it is no longer an intelligence we can fathom; we feel that it is infinitely above us: in a word, we feel that it is God Would that my feeble language might convey to you the full power of this impression; for until you are conscious of the infinite beauty and skill with which the numberless parts of nature have been fashioned and adjusted, you cannot appreciate the force of the conviction which the impression gives. You may make yourselves familiar with the dimensions of Mont Blanc; you may read the most glowing descriptions of this monarch of mountains, heightened by all the arts of eloquence or of poetry; you may cross the ocean and travel to the beautiful valley of Chamouni at its base; you may even climb its side, study its glaciers, and cross its fields of snow: but you will form no adequate conception of its grandeur, until, ascending one of the lofty mountain-peaks which surround it, you see its summit still towering above your head, apparently higher than before. So it is in the study of nature. No description can give you an adequate conception of the impression which it leaves upon the mind. It is not until, after long study, the student has become thoroughly acquainted with some one portion, however limited, of its wonderful economy, that he begins to appreciate the perfection of its parts, the infinite skill with which all have been adjusted, and the true grandeur of the whole.

By most men these heights of knowledge are unattainable. Why, then, should we hesitate to receive the evidence of a philosopher like Newton, who, after spending a long life in the investigation of nature, and with a success unparalleled in the history of science, uttered this memorable sentiment shortly before his death: "I do not know what I may appear to the world; but to myself I seem to have been only like a boy playing on the sea-shore, and devoting myself now and then to finding a smoother pebble or a prettier shell than ordinary, while the great ocean of truth lay all undiscovered before me." I know this sentiment has been so often repeated as to seem trite, but, coming from whom it does, it cannot be too often quoted. It is the testimony of the greatest master of science to its greatest and sublimest truth.

We can all recognize the marks of design in nature, and when we add to this evidence of our senses the testimony of a man like Newton, who assures us that the more our powers are enlarged, and the wider our knowledge becomes, the grander and vaster the design will appear, until it surpasses all our powers of thought or imagination, we begin to feel the full depth of the truth I have been endeavoring to enforce. If our minds are incapable of comprehending the plan, who could have been equal to the design? "Whence, then, cometh wis-

"dom, and where is the place of understanding, "seeing it is hid from the eyes of all living, and "kept close from the fowls of the air? God under- "standeth the way thereof, and he knoweth the "place thereof. For he looketh to the ends of the "earth, and seeth under the whole heaven, to make "the weight for the winds and a way for the light- "ning of the thunder. Then did he see it and de- "clare it; he prepared it, yea, and searched it out. "And unto man he said, Behold the fear of the Lord, "that is wisdom, and to depart from evil is under- "standing."

LECTURE III.

TESTIMONY OF OXYGEN.

WERE we to limit our regards to those physical qualities of the atmosphere which we studied in the first two Lectures, we should overlook the most wonderful adaptations in its divine economy. These properties belong to the atmosphere, in great measure at least, in virtue of its aeriform condition, and, so far as we know, an atmosphere composed of other gases, and still having the same density, would soften the intensity of the light, and diffuse the genial influences of the sun's heat, as well as air. Not so, however, with the chemical qualities of the atmosphere, which we are next to consider. These belong to the atmosphere solely as air, and could not have been obtained with any other known materials.

When a chemist wishes to investigate the nature of a new substance, his first step is to analyze it. Let us, therefore, as a preliminary to our present inquiry, ascertain what is the composition of this aeriform matter we call air. The air has been analyzed hundreds of times in every latitude and in every climate; and the result has been uniformly that which is given in the following table:—

Composition of air.

*Composition of the Atmosphere.**

Oxygen		20.61
Nitrogen		77.95
Carbonic Acid		.04
Aqueous Vapor	(average)	1.40
Nitric Acid	traces.	
Ammonia	traces.	
Carburetted Hydrogen	traces.	
		100.00

Composition in Tons.

Oxygen	1,233,010	billions of tons.
Nitrogen	3,994,593	" "
Carbonic Acid	5,287	" "
Aqueous Vapor	54,460	" "

Besides oxygen and nitrogen gases, which, as you will notice, are the chief constituents, there are always present in the atmosphere the vapor of water, carbonic acid, and ammonia gas ; and if we add to these uniform constituents the various exhalations constantly arising from the earth, we shall have as accurate an idea of the composition of the air as chemistry can give. While, however, the proportions of oxygen and nitrogen are almost absolutely constant, those of the other ingredients are very fluctuating, and the total quantity exceedingly small, never amounting in all, exclusive of aqueous vapor, to more than one part in a thousand, unless in some confined locality, and under very unusual circumstances. Do not, however, measure the im-

* Miller's Elements of Chemistry.

portance of these variable and in a degree accidental constituents by their amount, for, although present in such small quantities, they are not less essential in the atmosphere than the two gases which make up almost its entire mass.

Moreover, we must carefully avoid the error of considering air as a distinct substance, like water or coal. On the contrary, it is merely a mechanical mixture of its constituent gases, and is in no sense a definite chemical compound. Indeed, we may regard the globe as surrounded by at least three separate atmospheres, — one of oxygen, one of nitrogen, and one of aqueous vapor, — all existing simultaneously in the same space, yet each entirely distinct from the other two, and only very slightly influenced by their presence. To each of these atmospheres the Author of nature has assigned separate and different functions. They are like so many servants in a household, each with its distinct set of duties, which they discharge with a fidelity and diligence unknown to any earthly service. Let us consider what those duties are, and see how skilfully each is adapted to the offices which it is designed to fill.

Were all the other constituents of the air removed, the earth would still be surrounded by an atmosphere of oxygen, having about one fifth of the density, and exerting at the surface of the globe about one fifth of the pressure, of the present atmosphere. In studying the chemical relations of air, let us begin with some of the more important functions of this remarkable substance, and these will fully occupy our time during this and the next succeeding Lecture.

Properties of oxygen.

It is easy to prepare oxygen in a pure state. It is then a perfectly colorless and transparent gas, and so persistently does it retain its aeriform condition, that man has not been able, by the greatest pressure he can exert, or by the greatest degree of cold he can artificially produce, to reduce it to the liquid or solid state. A German chemist, Natterer, submitted this gas to the combined action of intense cold and a pressure of over forty-five thousand pounds, or twenty tons, on a square inch, but he did not succeed in changing its condition. This is quite an exceptional fact in nature. Almost all aeriform substances, when submitted to the action of great cold, are condensed to liquids, and many of these liquids have been frozen; but oxygen and nitrogen are an exception to the rule, and persistently retain their aeriform condition under the widest variations of temperature which exist on the earth. The importance of this fact will be seen at once on comparing the condition of the oxygen and nitrogen in the atmosphere with that of the aqueous vapor. A fall of temperature of only a few degrees will generally condense a portion of the vapor, and, small as is its relative amount, the resulting rain is at times poured down upon the earth in deluging floods;

Permanence of its aeriform condition.

and if you consider what must have been the destructive results had the whole mass of the atmosphere been liable to a similar fluctuation, even under extreme conditions, you will discover in the permanency of oxygen a most obvious adaptation of its properties to the thermal condition of our globe.

The permanently aeriform state of oxygen will appear still more remarkable when we consider how largely it enters into the composition of the solid crust of the earth. Oxygen belongs to that class of substances which the chemists call elements, because they have never succeeded in resolving them into simpler parts, and of all the elements it is by far the most widely diffused. As we have already seen, one fifth of the volume of the atmosphere consists of this gas; but this is a small amount compared with that which enters into the composition of most substances. You may be surprised at the statement, but it is nevertheless true, that between one half and one third of the crust of this globe and of the bodies of its inhabitants consists of oxygen. No less than eight ninths of all water is formed of the same gas. It makes up three fourths of our own bodies, not less than four fifths of every plant, and at least one half of the solid rocks. Remembering now that twenty tons of pressure on a square inch are not sufficient to reduce oxygen to a liquid condition, consider what must be the strength of that force which holds it thus imprisoned. Why, in this tumbler of water there are no less than six cubic feet of oxygen gas, condensed to a liquid condition, and held there by the continuous action of a force which can be measured only by hundreds of tons of pressure. We call the force chemical affinity; but who shall measure its power? Who but He who could make with such a subtile material the rocks, with which he "laid the foundations of the earth," and the waters which roll over its surface?

Oxygen gas, like all other forms of aeriform matter, tends to expand, and can be prevented from obeying this natural tendency only by enclosing it in an air-tight receiver. As it exists in our glass jars, under the ordinary conditions of temperature and pressure, one cubic foot of oxygen weighs 590.8 grains, although in its more expanded state, as it exists in the atmosphere at the surface of the globe, it has but one fifth of this density. One cubic foot of nitrogen gas weighs, under the same circumstances, 517.5 grains; but although there is such a decided difference between the specific gravities of the two gases, yet so perfectly are they mixed together throughout the whole extent of the atmosphere, that analysis has been unable to detect the slightest difference in composition between the air brought from the summits of the Alps and that from the deepest mine in Cornwall. Why, you may ask, do not these gases obey the well-known laws of hydrostatics, the heavier oxygen sinking to the surface of the earth, and the lighter nitrogen floating above it? Simply because gases, unlike the other forms of matter, have the property of *diffusing* through

Density.

Diffusive force.

each other, and existing together in the same space. The presence of one gas does not prevent the entrance of another into the space which it occupies, and if two open jars, containing different gases, are placed together, mouth to mouth, each gas will expand until it fills the whole volume. Moreover, the greater the difference between the densities of the gases, and the greater the consequent tendency to separate, the stronger is the force by which they are compelled to mix together. This force is called the force of diffusion, and it has been adjusted with great nicety in the plan of creation. Were this force in the atmosphere much less than it is, the two gases composing air would have separated partially, and the atmosphere have become unfitted for many of its important functions. Take, for example, the function of transmitting sound.

As the air is now constituted, there is a constancy of pitch, however far sound travels. Any tone once generated remains the same tone until it dies away. Its degree of loudness alters in proportion to the distance of the listener, but the pitch is constant. Were it not, however, for this law of diffusion, — were the atmosphere not perfectly homogeneous, and the gases of which it consists even partially separated, — there would have been a very different result. The constancy of pitch could no longer have been depended upon. The sound as it travelled would vary its pitch with the ever-varying medium through which it passed, and would arrive at the ear with a tone entirely different from that with which it started. Nor would it require any great difference in the medium to produce a sensible result, and to confuse all those delicate differences of pitch on which the whole art of music depends. Whenever, therefore, you may be next enjoying the grand Pastoral Symphony of Beethoven or the Requiem of Mozart, recall the careful adjustment of forces by which alone these magnificent creations of genius were rendered possible, and you cannot fail to recognize in this simple law of nature the same hand which first strung the lyre and made the soul of man responsive to its seven notes.

Passive and active conditions.

Returning again to the qualities of oxygen, let us notice, in the next place, that it is entirely destitute either of odor or of taste. This fact is a matter of common experience; for as oxygen exists in a free state in the atmosphere, it would there manifest these properties did they exist: and reflect how essential these negative qualities are to our comfort and well-being. Moreover, in its ordinary condition, oxygen seems entirely devoid of any active properties. It does not affect the most delicate and evanescent vegetable dyes, which the weakest chemical agents will either alter or destroy. And consider the oxygen as it exists in the air. How bland and seemingly inactive it is there! Reflect that it bathes the most delicate animal organism, that it pervades the minutest air-passages of the lungs, — remember that it is in contact with all matter, — and every substance will seem to bear evidence to the fact that oxygen in the state of gas possesses no active properties, and is incapable of manifesting any strong chemical force. And yet, if you infer that oxygen always appears in this passive condition, and is under all circumstances incapable of violent action, you will be entirely deceived; for so far from being one of the weakest, it is the strongest of the chemical elements, and beneath this apparent mildness there is concealed an energy so violent, that, when once thoroughly aroused, nothing can withstand it. A single spark of fire will change the whole character of this element, and what was before inert and passive becomes in an instant violent and irrepressible. The gentle breeze which was waving the corn and fanning the browsing herds, becomes the next moment a consuming fire, before which the works of man melt away into air.

Nature of fire.

And here I must correct an erroneous, although very common impression, that there is something substantial in fire. This is one of those ideas, originating in an illusion of the senses, which we have inherited from a more ignorant age, and which our modern science cannot wholly dispel from the popular mind. Fire was formerly regarded as one of the elementary forms of matter, and all burning was supposed to consist in the escape of this principle of fire, previously pent up in the combustible substance. In support of this doctrine the old philosophers confidently pointed at flame as the visible manifestation of the escaping fire-element. And, childish as this doctrine may seem, it was the prevalent belief of the world for at least two thousand years.

The last phase which this doctrine assumed was the phlogiston theory of the last century. In the hands of Bergmann and Stahl, the vague ideas of the time received a more material form, and were embodied in a philosophical system. They termed the principle of fire *phlogiston*, and burning, or the escape of fire, *dephlogistication*, and their ingenious system did not a little to retard the progress of truth. It never seems to have occurred to the philosophers of that age to weigh the results of burning, or if they did by accident, they most blindly ignored the result, and it was not until 1783, when the true theory of combustion was discovered, that modern chemistry began. The fortunate discoverer was Lavoisier. He proved, by simply weighing the products of combustion, that burning, instead of being a loss of phlogiston, is a union of the burning substance with the oxygen of the air, and this theory is now one of the best established principles of science.

Burning is merely chemical change, and all combustion with which we are familiar in common life is a chemical combination of the burning substance, whether it be coal, wood, oil, or gas, with the oxygen of the air. Combustion is simply a process of chemical combination, and the light and heat which are evolved in the process are only the concomitants of the chemical change. Why those mysterious influences of light and heat are radiated from the coal which is combining with oxygen in our grates, we may understand better hereafter; but this much we already know, — that the sensations of light and heat are caused by waves of an ethereal medium breaking upon the extremities of the delicate nerves of our human organism; and that such waves are set in motion during the chemical change which we call combustion. What the chemist chiefly studies, however, is the change itself, and to this we will for the present confine our attention.

The products of ordinary combustion, that is, the compounds of oxygen with the elements of coal, wood, and gas, are only two in number, carbonic-acid gas, and aqueous vapor. These, as is well known, are perfectly colorless and transparent aeriform substances, wholly without odor or taste, and entirely devoid of every active quality. For this reason they escape without observation from the burning wood, ascend our chimneys, and by the force of diffusion are spread throughout the atmosphere; but if, as may readily be done by chemical means, we collect the neglected smoke and weigh it, we shall find that it weighs much more than the burnt wood, and, as more careful experiments will show, its weight is exactly equal to that of the wood added to that of the oxygen of the air consumed during the burning.

And, my friends, this smoke, though so long unnoticed by man, was not overlooked by the

Author of nature. It is a part of his grand and beneficent design in the scheme of organic nature. No sooner do the products of that wood burning on your hearth escape into the free expanse of the outer air, than a new cycle of changes begins. The carbonic acid and the aqueous vapor, after roving at liberty for a time, are absorbed by the leaves of some wide-spreading tree, smiling in the sunshine, and in the tiny laboratory of their green cells are worked up by those wonderful agents, the sun-rays, into new wood, absorbing from the sun a fresh supply of power, which is destined, perhaps, to shed warmth and light around the fireside of a future generation.

But let us not anticipate our subject. In a future Lecture I hope to discuss this wonderful cycle of changes at some length. At present I wish to direct your attention to the remarkable contrast of qualities presented by the element oxygen in its active and passive conditions. How is this complete inversion of properties to be explained? There is a cloud of mystery hanging over the subject, which the progress of knowledge has not as yet entirely dispelled; but so far as the cause is known, I will endeavor to make it intelligible. The difference in the action of oxygen in these two conditions depends on temperature. At the ordinary temperature of the air its chemical affinities are dormant, and, although endowed with forces which are irresistible when in action, it awaits the necessary conditions to call them forth. One of the grandest works of ancient art which have come down to us is the colossal statue of the Farnese Hercules. The hero of ancient mythology is represented in an erect form, leaning on his club, and ready for action; but at the moment every one of the well-developed muscles of his ponderous frame is fully relaxed, and the figure is a perfect ideal of repose, yet a wonderful embodiment of power. Here in this antique we have most perfectly typified the passive condition of oxygen, the hero of the chemical elements. Raise now the temperature to a red heat, and in a moment all is changed. The dormant energies of its mighty powers are aroused, and it rushes into combination with all combustible matter, surrounded by those glorious manifestations of light and heat which every conflagration presents.

Fire, how sustained.

In order to evoke the latent forces in the oxygen of the atmosphere, it is not necessary, however, to raise the temperature of any considerable portion either of the gas or of the combustible. There is a provision in nature by which chemical combination, once started at any portion of the combustible mass, is sustained until the whole is consumed. All chemical combination is attended by the evolution of heat, and in the combination of oxygen with most combustible substances the amount of heat thus generated is so great, that by the burning of one portion sufficient heat is evolved to raise the temperature of a second portion to the point of ignition, and thus the process is continued. Consider, for example, what takes place in the burning of a jet of gas. We start the combustion by bringing the flame of a lighted match over the orifice. By this the temperature of the gas and that of the air surrounding it are raised to a red-heat, and chemical combination at once ensues. But the chemical union, as just stated, is attended with the evolution of great heat, which, before it is dissipated by the air, raises to the point of ignition the temperature of the next portion of gas issuing from the burner. This, combining in its turn with the oxygen of the air, generates a fresh quantity of heat, and thus keeps up the combustion so long as the gas is supplied. What I have shown to be true of these gas-burners is equally true of all ordinary combustion, and so a single spark may be sufficient to light up a conflagration which will reduce to ashes a whole village or involve a city in ruin.

Thus it appears that burning is chemical combination with oxygen, that this union is attended with the evolution of heat, and that a high temperature is the condition under which oxygen manifests its latent power. But you may say, these facts do not explain the difference between the two states of oxygen, they merely give the conditions under which these states are manifested; and this is true. Why it is that at one temperature oxygen is so completely passive, and at another temperature, a few hundred degrees higher, so highly active, we not only cannot explain, but we cannot even conjecture. The facts are undisputed; the causes lie hid among the counsels of the All-Wise.

The temperature at which oxygen assumes its active condition is called the point of ignition. Although fixed for each substance, it differs very greatly with the different kinds of combustible matter, being determined, apparently, by their relative affinities for the great fire element. Thus phosphorus ignites at a temperature less than that of boiling water, sulphur at 500°, wood only at a full red heat, anthracite coal at a white heat, while iron requires the highest heat of a blacksmith's forge. Beginning with a phosphorus match, which can be ignited by friction, and using the more combustible materials as kindlings, we can readily attain in our furnaces the highest temperature required, and thus the energies of this powerful agent are fully at the command of man. But notice at the same time that the point of ignition of wood, coal, and common combustibles, has been placed sufficiently above the ordinary temperature of the air to insure the general safety of our combustible dwellings; and when we consider how liable they are, even now, to accidents from fire, we shall appreciate the care which has been taken by our Heavenly Father to guard us against this terrible danger.

Point of ignition.

But even this precaution would have been insufficient to secure safety, were it not that the active energies of oxygen, even when aroused, have been most carefully tempered by extreme dilution.

It would be easy to show by experiment that the slowness of combustion depends on the fact that in the atmosphere oxygen is mixed with a great mass of an inert gas, and the proportions have been so adjusted in the scheme of creation as generally to restrain the awakened energies of the fire-element within the narrow limits which man appoints; but when, through his misfortune or carelessness, it overrides these limits, and, from administering to man's wants, becomes the agent of his destruction, we are reminded in the awful conflagration by what a delicate tenure we hold our earthly possessions, and how small a change would be sufficient to involve all organized matter in a general conflagration. Remember now that fire is one of the most valuable servants of mankind; that it is the source of all artificial heat and light; that in the steam-engine it is the apparent origin of that power which animates the commerce and the industry of the civilized world; that under its influence iron becomes plastic, and the ores give up their metallic treasures; that it is, in fine, the agent of all the arts, — and you cannot wonder that in a ruder age the Romans should have enthroned its presiding deity on Olympus, or the Persians worshipped its supposed essence as divinity itself. Looking at it again, in the light of modern science, as merely the manifestation of the latent power of this bland and diffusive atmosphere, the truth seems almost incredible. To think that this, the strongest of the chemical elements, which, although a permanent gas, forms more than one half of the solid earth, and is endowed with such mighty affinities that it is retained securely in this solid state, could have been so shorn of its energies as not to singe the down of the gossamer, and yet so tempered that its powers may be evoked at the will of man and made subservient to his wants! To me the double condition of oxygen is one of the most remarkable phenomena of nature. I ponder over it again and again, with increasing wonder and admiration at the skill of the infinite Designer, who has been able to unite in the same element perfect mildness and immeasurable power. It seems as if the millennium of the Hebrew prophet were prefigured in the atmosphere. "The wolf also shall dwell with the lamb, and the "leopard shall lie down with the kid, and the calf "and the young lion and the fatling together, and "a little child shall lead them."

If I have succeeded in making clear the relations of this twofold character of oxygen to man and his works, I think that you cannot fail to have been impressed with the evidence of design which the subject affords. This evidence is seen in the facts, first, that the same element is at different temperatures endowed with such opposite and apparently incompatible qualities; secondly, that in each of its conditions the properties are so skilfully adapted to the functions which it is appointed to perform; thirdly, that the temperature at which it assumes its active state has been so accurately adjusted to the thermal conditions of the globe; and lastly, that its active energies have been so carefully guarded, and placed to so great a degree under the control of man. But we have not as yet half exhausted the subject. Here, as everywhere else in nature, the argument is cumulative; the more we study, and the more our knowledge is enlarged, the more it grows upon us; and wherever we may leave the field, we always are conscious that there is a still richer harvest to be reaped beyond.

Oxygen the chief world-builder.

If the crust of the globe is a fair sample of the whole mass, oxygen was the chief material employed by the Great Architect in constructing our earth. Moreover, world-building was a process of burning like those we have been studying, and the foundations of the earth were undoubtedly laid in flames.

When we attempt to break up the various materials around us into simpler parts, we soon reach a class of substances which cannot be further decomposed. Simple inspection will show that granite rock, for example, is a mixture of three minerals, called feldspar, mica, and quartz. We know, also, that feldspar consists of alumina, potash, and silica, that mica contains the same materials in different proportions, and that quartz is silica alone. Lastly, the chemists have discovered that alumina is composed of *aluminum* and *oxygen*, potash of *potassium* and *oxygen*, and silica of *silicon* and *oxygen*. But here we must stop; for when you ask us of what these last-named materials are made, we find ourselves in the condition of the old philosopher, who got on very well with his flat earth, supporting it on an elephant, and the elephant on four tortoises, until he came to seek a resting-place for the tortoises; but then his theory failed. So is it with our science. These undecomposed materials are the blocks on which the whole is built; and we are totally ignorant of what lies below.

We call all substances which have never yet been decomposed, whatever may be their nature, chemical elements, and of such about sixty-five are now known. Setting apart oxygen as the supporter of combustion, the great mass of the remaining elements are combustible; that is, under certain conditions they combine rapidly with oxygen, evolving light and heat. Indeed, many of the combustible substances with which we are most familiar are elements. Charcoal is an element, phosphorus is an element, sulphur is an element, iron and all other metals are elements, and out of such combustible materials, together with oxygen, the world is made, and chiefly out of oxygen.

When we burn charcoal in air, or pure oxygen gas, the burning is a process of world-making. The charcoal combines with oxygen, and the result is a transparent, colorless gas, called carbonic-acid. Many may not have heard of such a substance before, but it is always present in the atmosphere, at least in small quantities, and if we continue our process of world-making a little further, we shall find that it enters into the composition of some of the most familiar rocks and minerals.

I have at the bottom of this closed glass tube a small piece of a yellowish-white metal, looking very much like a flattened shot; and so it is, but the metal is not lead, although it resembles lead very closely. Like lead it is quite soft, and can be easily beaten into leaves thinner than writing-paper; but it is very much lighter than lead, and tarnishes so rapidly in the air that we are obliged to keep it thus protected. We call the metal calcium, and although you may never have seen the substance before, it is one of the most abundant metals in nature; yet a great rarity, because of the extreme difficulty with which it is extracted from its ores. When heated to redness, calcium burns with a brilliant white light and a scintillating flame. In burning, it combines of course with oxygen, and the result is lime, common quicklime, such as is used for making mortar. This is a process which in the original world-making must have played a very important part, for lime rocks form a large portion of the earth's crust. None of these rocks, however, will slake like quicklime, and we must go a step further in our world-building, and bring in the agency of water, before we can reach the actual condition of things.

We have now before us two products of burning, one a solid, called lime, made by uniting calcium with oxygen, the other a gas, called carbonic-acid, made by uniting charcoal with oxygen. Both are soluble to a certain extent in water, and these clear solutions, called lime-water and soda-water respectively, are even more familiar to you than the substances themselves. Mix now the solutions together. The water becomes at once very turbid, and there soon settles from it a white powder. The lime and carbonic acid have united, and this is the result. If we collect and examine the white powder we shall find that it is chalk, and from the same material, spread in thick layers over the ocean-bed, and subsequently hardened by the mutual action of heat and water, have been formed limestone, marble, and the different varieties of lime rock, which are all ores of calcium.

But we may study with profit a second example of world-building. I have here a small quantity of another very abundant element, called silicon, but, like calcium, a comparative rarity, because it is with difficulty obtained pure. It resembles in many respects carbon, and has been observed in three different states, corresponding to charcoal, graphite, and diamond. Like carbon, it also is combustible, combining with the oxygen of the air when heated to a high temperature, and forming a very hard white solid, called by chemists silica, which is the same thing as quartz, rock-crystal, agate, jasper, calcedony, opal, &c. All these familiar minerals are merely different conditions of this one material, and contain over one half their weight of oxygen gas. When ground to a coarse powder by the action of running streams, they become sand, and the grains of sand, compacted together, form sandstone and similar rocks; and you will begin to appreciate the enormous amount of silicon which must have been burnt up in the process of world-making, when you learn that at least one half of the solid crust of the earth consists of silica in its different varieties.

Setting aside the silica for a moment, let us turn to another very widely distributed element, called aluminum. This brilliant white metal, comparing favorably even with silver in lustre, was until very recently as great a rarity as calcium or silicon; but within a few years a process has been invented by which it can be extracted from its ore at a cost sufficiently low to render the metal available in the arts, and it has now come into quite general use for making mathematical instruments, for jewelry, and for similar purposes. It forms also, with copper, a valuable alloy, which does not readily tarnish, and resembles gold so closely that the two cannot be distinguished by their external appearance.

Aluminum, like most of the metals, is combustible, although it does not burn readily in the air, unless the temperature is very high and the metal finely subdivided; but it then burns very brilliantly, emitting a vivid light, and forming a compound called by chemists alumina, which is melted by the intense heat to a yellowish transparent glass, and is the same substance from which nature makes the sapphire and the ruby. Emery also, which, on account of its great hardness, is used so largely for polishing, is only a rougher form of the same material. Unite now the alumina to silica, and we get clay. Burn the clay, and we have, according to the fineness of the materials, porcelain, pottery-ware, or bricks.

Taking next the element magnesium, which is also a brilliant white metal, allied to zinc, you notice that it takes fire even in the flame of a candle, and burns with dazzling brilliancy. The result is magnesia, so much used as a medicine. Unite magnesia to silica, and we have, according to the proportions, hornblende or augite, two minerals which abound in many varieties of rock. Add water to the composition, and we get also serpentine or soapstone, with several other allied mineral species.

I might multiply these illustrations indefinitely, but I will limit myself to only one other example. Here is a metallic element called potassium, so light and combustible that it swims and burns on water. Burning in water may seem, at first sight, very paradoxical; but in studying chemistry, we must be ready to give up old prejudices. Water is almost pure oxygen, containing in the same volume more than one hundred times as much of the fire-element as air, and all combustibles would burn in water were it not that the oxygen is imprisoned in the liquid by an immensely strong force. Potassium, however, has such intense chemical affinities that it will break through all bars and bolts in order to unite with oxygen, and it therefore burns thus brilliantly even in the midst of water. The result is a white solid called potash, which in the experiment dissolves in the liquid. Melt together now potash, lime, and silicious sand, and we have glass. Unite silica, alumina, and potash, and we get feldspar; combine them in little different proportions, and we have mica; varying again the proportions, we obtain garnet. Lastly, mix quartz and feldspar together with mica or hornblende, in an indiscriminate jumble, and we have the several varieties of granitic rocks.

Such then are some of the steps in the process of world-building. I do not mean to imply that we can reproduce all these substances in our laboratories, although even this is true in almost every case. My object is only to show what must have been in general the process of nature, and to make evident the fact that oxygen has been the chief world-builder. But why call oxygen the world-builder more than the other elements? This diagram will answer, and it illustrates one of the most

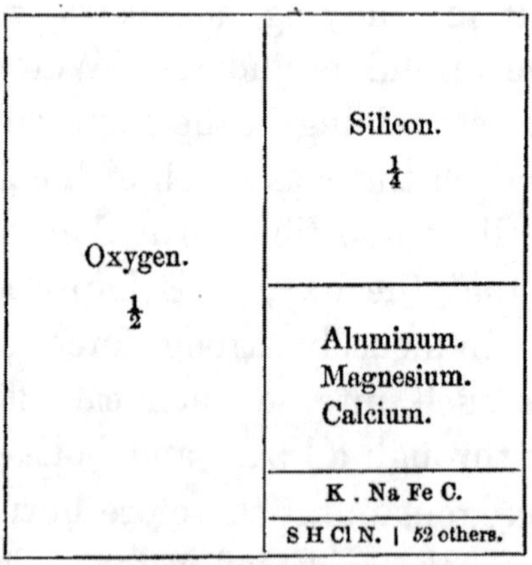

remarkable facts to which the study of this function of oxygen has led. Of the 65 known elements, thirteen alone make up at least $\frac{99}{100}$ of the whole known mass of the earth. Of this, oxygen forms about $\frac{1}{2}$, silicon about $\frac{1}{4}$; then we have aluminum, magnesium, calcium, potassium (K), sodium (Na), iron (Fe), carbon (C), sulphur (S), hydrogen (H), chlorine (Cl), and nitrogen (N) filling up nearly the other fourth, while the 52 remaining elements — including all the useful metals except iron — do not constitute altogether more than $\frac{1}{100}$. The diagram, however, only represents the relative proportions very rudely, as the subdivisions are necessarily based on very rough estimates and imperfect data.

Evidently, then, so far as our knowledge extends, oxygen, silicon, and carbon, together with a few metals, have been the chief building-materials employed by the Great Architect, and oxygen has been, as it were, the universal cement by which the other elements have been joined together to form that grand and diversified whole we call our earth.

One more remark in regard to this subject, and I will close my Lecture. It is probable that there was a time, long before even geological history, when the elements were in a free state; when the oxygen now solidified was a gas forming an atmosphere, and when at the fiat of the Almighty the union of the elements began. Then our earth became a bright, burning star, radiating its heat and light into space. Indeed, if we accept the nebular hypothesis of Laplace, the earth was formerly a part of the sun, was thrown off by the centrifugal force from his burning mass, and like a spark from a forge soon burnt out, although after this lapse of time the great central fire is burning still. But whether Laplace is right or not, this much is certain. The crust of the earth, so far as we can examine it, is like a burnt cinder, and the atmosphere of oxygen which surrounds it is merely the residuum left after the general conflagration,—left because there was nothing more to burn. Unmeasured ages have passed away since then; the earth's crust has cooled and solidified; the waters have been condensed and gathered into the great ocean-basins; the dry land has been covered with verdure and peopled with all kinds of four-footed beasts, winged fowls, and creeping things; the waters have been tenanted with countless forms of swimming creatures; and, last of all, man has come to live in this fair creation, and study the wonders of his dwelling-place. He finds on the earth's burnt crust an abundant supply of combustible material for all his wants. But if the world was once burnt up and the elements glowed with fervent heat, how is it that these combustibles have been left unconsumed? Modern science has been able to answer this question. It has discovered that during the long geological ages a silent power has been slowly recovering a small amount of combustible material from the wreck of the first conflagration. The sunbeam has partly undone the work of the fire, and whatever now exists on the earth unburnt, wood, coal, or metal, we owe to that wonderful agent, the solar light. How the result has been accomplished, I propose to consider in a future Lecture.

LECTURE IV.

TESTIMONY OF OXYGEN.— *Concluded.*

Besides the two extreme conditions of oxygen, there exists still a third, in a measure intermediate between them, but still differing essentially from either,—a condition in which the element discharges functions, less brilliant it is true, but not less interesting and instructive, than those which we studied in the last Lecture. The phenomena in which this condition of oxygen is chiefly active require, as a general rule, months, or even years, for their full manifestation. Moreover, they are so silent and unobtrusive, as frequently to be passed unnoticed; but nevertheless, when we have become acquainted with their magnitude and importance, I am sure you will agree with me that they far surpass in true grandeur those dazzling displays of power which the fire-element manifests when fully aroused. This third phase of the element can be best studied in its effects, and to two of these I will now ask your attention.

Third condition of oxygen.

Every one knows that, when wood or any other organized structure is exposed to the moist atmosphere, it gradually decays. It first becomes rotten, and then slowly disappears.

Decay.

All may not know, however, that decay consists in a slow union of the organized structure with oxygen, and that the log of wood which is left to rot in the forest undergoes precisely the same change as one which is burnt on the hearth. The sole difference is, that, while the last is burned up in a few hours, the first entirely disappears only after the lapse of many years. Wood, like all organized vegetable structures, consists mainly of three elements, charcoal, hydrogen, and oxygen. When heated on the hearth, in contact with the air, it takes fire and burns; that is, its combustible elements combine with oxygen, the carbon to form carbonic acid, and the hydrogen to form aqueous vapor, both of which escape by the chimney. But of these two ingredients of the wood, hydrogen is by far the most combustible; that is, it has the greatest tendency to combine with oxygen, and therefore burns first, leaving the less combustible charcoal in the form of glowing coals. If at this point we take up one of these coals and quench it in water, it will be found to be common black charcoal; but if left on the hearth, the coal also burns, gradually smouldering away, and passing up the chimney as carbonic-acid gas.

Quite a similar succession of phenomena is presented in the forest during the process of decay. In decay, as in burning, the oxygen of the air unites with the hydrogen of the wood more rapidly than with the charcoal, and in consequence the rotten wood becomes darker and darker, from the excess of black charcoal, as the change advances. Moreover, if the supply of air is insufficient, as when the wood is buried in swamps, it is finally reduced to coal, which corresponds to half-burnt wood. In the open air, however, the charcoal as well as the hydrogen is burnt, and the log of wood is resolved, as in ordinary combustion, into carbonic acid and water, leaving only a few handfuls of earth to mark the spot where it lay. This change requires years before it is fully consummated, and it is not therefore wonderful that its nature should not have been understood until a comparatively recent period. Thanks to modern chemistry, the subject is now less obscure. We may not be able to trace all the steps of the process, but this much we know. Decay and burning are essentially the same chemical change. The substances involved are the same, the results are the same, and we have even been able to prove that the amount of heat generated is the same, the only difference being, that, in burning, the whole amount of heat is set free in a few hours, producing phenomena of intense ignition; while in the process of decay the same quantity, slowly evolved during perhaps a century, escapes notice.

It has been observed, that, if wood be left in contact with dry oxygen, it may be kept indefinitely without undergoing change, — a fact sufficiently proved by the mummy cases of Egypt, which in that dry climate have been preserved for over three thousand years; — also, that if wood is impregnated with many salts, as in the process of Kyanizing or Burnetizing, decay may be arrested even in a damp situation for a long time. In both cases the prevention depends on destroying certain very unstable compounds which are present in all green wood, and which start the decay. These are termed by chemists albuminous substances, the chief of which, vegetable albumen, is almost identical with the white of an egg. The great bulk of all *vegetable* structures, as was stated above, consists of only three elements, charcoal, hydrogen, and oxygen; but these albuminous substances — which, as a rule, are present only in very small quantities — contain in addition to the three just mentioned a fourth element, nitrogen. Partly because they contain nitrogen, and partly, unquestionably, in consequence of the complex manner in which the four elements are combined, the albuminous substances are vastly more unstable than the great mass of vegetable matter, and in the presence of moisture they soon undergo an internal change, called putrefaction, or fermentation, by which they are broken up into simpler compounds. The precise nature of the process is not understood, but nothing appears to be added to the substance unless it be water, and the change seems to consist in the falling to pieces of a complex organic structure. At all events, oxygen gas is not essential to the process, but the oxygen of the air which happens to be in contact with the fermenting substances in some mysterious way undergoes a remarkable change. It becomes endowed with active properties even at the ordinary temperature, and, with its affinities thus exalted, slowly consumes the wood, together with all other organic compounds present. Moreover, the process, once started, sustains itself. As, in burning, the union of the combustible matter with oxygen engenders sufficient heat to maintain the surrounding gas in its highly active modification, so in like manner the process of decay seems to modify continually the neighboring oxygen, arousing its energies, and thus continuing the change when once begun.

While the plant is in great measure made up of non-nitrogenized substances, the animal, on the other hand, consists almost entirely of albuminous compounds. The flesh, the nerves, and the bones of our bodies all contain nitrogen, and, like the vegetable albumen, are prone to decay; and this change is constantly going on in our living members. In a most profound sense, "in the midst of life we are in death." The materials of our bodies are being constantly renewed, and the great mass of their structure changes in less than a year.* At every motion of your arm, and at every breath you draw, a portion of the muscles concerned is actually burnt up in the effort. During life, in

* The rapidity of the change has not been accurately determined. Some authors state that the great mass of the body changes every month, and when we consider the large quantities of water, carbonic acid, and ammonia daily secreted, the statement appears credible; but in the absence of direct proof we have set the limit unnecessarily high in order to avoid the slightest exaggeration.

some utterly mysterious manner, beyond the range of all human science, the various gases and vapors of the atmosphere, together with a small amount of a few earthy salts, are elaborated into various organized structures. They first pass into the organism of the plant, and thence are transferred to the body of the animal; but no sooner are they firmly built into the animal tissues, than a destructive change begins, by which before long they are restored to the air or the soil, only to renew the same cycle of ceaseless change. Life, during its whole existence, is an untiring builder, the oxygen of the atmosphere a fell destroyer; and when at last the builders cease, then the spirit takes its heavenward flight, and leaves the frail tenement to its appointed end. Dust returns to the dust, and these mortal mists and vapors to the air.

I know that there are some who entertain a vague fear that these well-established facts of chemistry conflict with one of the most cherished doctrines of the Christian faith; but so far from this, I find that they elucidate and confirm it. I admit that they do disprove that interpretation frequently given to the doctrine of the resurrection, which assumes that these same material atoms will form parts of our celestial bodies; but then I find that this interpretation is as much opposed to Scripture as to science. The Saviour himself, in his reply to the incredulous Sadducees, severely rebuked such a material con-

Doctrine of the resurrection.

ception of his spiritual revelation, and the great Apostle to the Gentiles, in his vision of the glorified body, distinctly declares that this body is not the body that shall be; but that, as the grain sown in the furrow rises into the glory of the full-eared corn, "so when this corruptible shall have put on incorruption, and this mortal shall have put on immortality," our natural body, sown in dishonor and weakness, will be raised a spiritual body, clothed in glory and in power. "And as "we have borne the image of the earthy, we shall "also bear the image of the heavenly."

The glorious doctrine of the resurrection here presented, modern scientific discoveries most fully confirm. They have shown that our only abiding substance is merely the passing shadow of our outward form, that these bones and muscles are dying within us every day, that our whole life is an unceasing metempsychosis, and that the final death is but one phase of the perpetual change. Thus the idea of a spiritual body becomes not only a possible conception, but, more than this, it harmonizes with the whole order of nature; and now that we can better trace the processes of growth in the organic world, and understand more of their hidden secrets, the inspired words of Paul have acquired fresh power, and convey to us a deeper meaning than they ever gave to the early Fathers of the Church. It is no wonder that, when men were less enlightened, the doctrine should have been misinterpreted; but now, when the truth has been illuminated by the study of nature, why longer harass the understanding and vex the spirit with these material clogs? Hear again the words of the Apostle: "This I say, brethren, that flesh and blood cannot inherit the kingdom of God; neither doth corruption inherit incorruption." "For this corruptible must put on incorruption, and this mortal must put on immortality." And now, turning to the glorious truth as Christ revealed it and Paul preached it, how greatly is our faith strengthened by these lights of nature! All philosophy assures us that the finite and limited can be manifested only under form.

> "That each, who seems a separate whole,
> Should move his rounds, and, fusing all
> The skirts of self again, should fall
> Remerging in the general Soul,
>
> "Is faith as vague as all unsweet:
> Eternal form shall still divide
> The eternal soul from all beside;
> And I shall know him when we meet."

Chemistry has shown us that it is the form alone of our mortal bodies which is permanent, and that we retain our personality under constant change; and lastly, in organic nature, the sprouting of the seed, the breaking of the bird from the egg, the bursting of the butterfly from the chrysalis, and ten thousand other transmutations not less wonderful, which we are daily witnessing around us, all unite their analogy to elucidate and confirm the glorious and comforting doctrine of a material resurrection in form.

Moreover, when we remember that our organs of vision and hearing are capable of receiving impressions either of light or sound only when the rapidity of the undulations which cause them is comprised within certain very narrow limits, and when we recall the facts stated in a previous Lecture, that there are waves of light and sound of which our dull senses take no cognizance, that there is a great difference even in human perceptivity, and that some men, more gifted than others, can see colors or hear sounds which are invisible or inaudible to the great bulk of mankind, you will appreciate how possible it is that there may be a world of spiritual existence around us — inhabiting this same globe, enjoying this same nature — of which we have no perception; that in fact the wonders of the New Jerusalem may be in our midst, and the songs of the angelic hosts filling the air with their celestial harmony, although unheard and unseen by us. Let me not be understood as implying that science has in any sense revealed to us a spiritual world, or that it gives the slightest shadow of support to those products of imposture, credulity, and superstition, which, under the name of witchcraft, mesmerism, or spiritualism, have in every age of the world deceived so many. The only revelation man has received of a spiritual existence is contained in the Bible; but modern science has rendered the conception of such an existence possible, and in this way has removed a source of doubt. The materialist can no longer say that the spiritual world is inconceivable; for these discoveries show that it may be included in the very scheme of nature in which we live, and thus, although science may not remove the veil, it at least answers this cavil of materialism.

Returning now to the main subject, consider for a moment the importance of this ghostly office of oxygen in the scheme of organic nature. Reflect how soon this fair world would become a great charnel-house were it not for these provisions, by which its youth is constantly renewed. Remember also that this process of decay furnishes the materials from which young life builds her fresh and blooming forms; that, although in the midst of life we are in death, it is equally true that death is only a phase of life. Then these changes of outward nature will assume a new aspect. It will be seen that they are the beneficent provisions of infinite wisdom, in themselves full of interest and beauty, and only sad and melancholy as they are associated with bereaved affections and disappointed hopes, or with that only real death, the moral death of the soul. "O death, where is thy sting? O "grave, where is thy victory? The sting of death "is sin; and the strength of sin is the law. But "thanks be to God, which giveth us the victory "through our Lord Jesus Christ."

I might profitably occupy several hours in describing the various processes of slow combustion, for they are all rich in illustrations of skilful design; but I must content myself with only one other example, and from the many which crowd upon me I have chosen respiration, because it is so well understood and because it is so intimately associated with our own physical existence. Respiration is a true example of combustion. The seat of the combustion is the lungs. The substance burnt is sugar. The products are carbonic-acid gas and water.

Respiration.

The materials of animal food may be divided into three classes: non-nitrogenized substances, such as starch and sugar; nitrogenized substances, like lean meat and eggs; and, lastly, fatty substances, like butter. To these must be added a small proportion of earthy salts, which, however, as they enter into the composition of almost all varieties of food, do not properly form a distinct class. All of the three classes of food are absolutely necessary to support the life of the higher animals, and especially of man, and they are all contained in those articles of diet which will of themselves alone sustain life. Milk may be regarded as the type of animal food.

*Composition of Milk.**

	Natural state.	Evaporated to dryness.
Water,	87	
Curd or casein,	4½	34¾
Butter or fat,	3	23¾
Sugar (of milk)	4¾	37
Ash (nearly)	¾	4½
	100	100

* Johnston's Chemistry of Common Life.

It contains, in the first place, a non-nitrogenized substance, sugar; in the second place, a nitrogenized substance, casein, which separated from milk forms cheese; and, lastly, a fatty substance, which when separated by churning forms butter.

	Wheaten Bread.	Lean Beef.
Water (and blood),	45	78
Fibrin or gluten,	6	19
Fat,	1	3
Starch, &c.,	48	
	100	100

Bread, again, consists of starch, a non-nitrogenized substance, of gluten, a nitrogenized substance, and it also contains about two per cent of a peculiar oil. No article of food which does not contain all three of these classes of substances can alone support life for any length of time. A man would starve to death on starch alone, on meat alone, or on butter alone. The relative proportion, however, in which these three classes of substances are required by man, depends on his outward circumstances, such as the climate, his physical activity, his occupation, or his peculiar temperament, and to the right balance of his food he is guided by experience.

The different classes of food serve different functions in the body. The nitrogenized and a portion of the fatty substances are used to supply the constant waste of the tissues which result from all the animal processes. They are in some unknown way vitalized in the system, and converted into new muscles, tendons, and nerves, which take the place of those that have been used up. On the other hand, the non-nitrogenized substances, such as starch, are supposed to take no part in the formation of new tissues, and to be merely the fuel by which the animal heat is maintained. Let us very briefly follow these substances through the body, and see when and how they are burnt.

By far the greater part of our daily food consists of varieties of starch or sugar. These two substances are almost identical in composition, and starch may be converted into sugar with the greatest ease. Leaving out of view the large amount of water which all our food contains, we find that of wheaten bread no less than 39 per cent consists of starch or sugar; of potatoes fully 92 per cent is made up of the same materials, and in general they form over four fifths of the solid part of all our food. These substances when taken into the stomach are almost instantaneously converted by the saliva and the gastric juice into the variety of sugar known as grape-sugar, so called because it is the sweet principle of ripe grapes. The sweet principle of honey and molasses, and the incrustation which is so frequently seen on figs and raisins, are also essentially the same substance. Grape-sugar, being very soluble, dissolves in the water present, and the solution is absorbed by the veins which ramify on the surface of the intestinal canal, into which the digested food passes from the stomach. The blood, now containing sugar in solution, returns through the liver to the right side of the heart, and by this organ, which consists essentially of two ingeniously contrived force-pumps, arranged side by side, it is forced through the lungs, where the sugar is brought in contact with the air. Let us next examine for a moment this remarkable structure.

The lungs, as is well known, consist of two large organs, on either side of the chest, called the left and the right lung. The right lung is divided into three smaller lungs, called lobes, the left into but two. On examining any one of these lobes it will be found to be made up of an immense number of small membranous bags, all closely packed together. These small bags, called cells, connect by means of the bronchial tubes and windpipe with the air, through the nose and mouth. They vary in size, but on an average are about $\frac{1}{100}$ of an inch in diameter, and the total number of the cells in the lungs has been estimated at six hundred millions. Their walls are exceedingly thin, and the cells may therefore be easily compressed. The whole mass of the lungs is also exceedingly elastic, and by the action of a system of muscles their volume is alternately increased and diminished in the process of respiration. The amount of air which is thus drawn into the cells, and again expelled at each inspiration, differs in different individuals. The average quantity in the ordinary tranquil respiration of an adult is about a pint; but in a full respiration it may be as much as two and a half pints, and by an effort the lungs may be made to inhale from five to seven pints. As the average in health is about eighteen inspirations a minute, which corresponds to about eighteen pints of air inhaled and exhaled, it follows that three thousand gallons of air pass through the lungs of an adult man every day. Some estimate it as high as four thousand gallons a day for an average man in average circumstances, and as high as five thousand seven hundred gallons a day for an athletic man undergoing severe exertion. In order that you may form an idea of this quantity, I will add that four thousand gallons of air would fill a room measuring about eight and a half feet in each dimension.

Let us now turn to the blood, and examine the apparatus by which it is exposed to the air in the lungs. As we have already seen, the blood charged with sugar is received into the heart, from whence it is pumped through a long tube, called the pulmonary artery, into the lungs. This artery divides again and again until it is reduced to very small capillary tubes, which ramify on the surfaces of the air-cells. The walls of these capillaries are formed of the thinnest conceivable membrane, so as to bring the blood into as close contact as possible with the air. Here oxygen gas is absorbed in large quantities, and carbonic-acid gas evolved. The blood now holds in solution at the same time oxygen gas and sugar, and, thus charged, it returns by a series of veins to the left side of the heart, when by the second of the two force-pumps it is again forced through the general circulation of the body. In the mean time the oxygen absorbed by the blood while in the lungs burns up the sugar. Sugar, like wood, consists of carbon, hydrogen, and oxygen. The last two are present in the proportions to form water, so that sugar may be said to be composed of charcoal and water. Of these two substances the charcoal only is combustible. This, during the circulation of the blood, is slowly burnt up by the dissolved oxygen, and converted into carbonic acid, which remains in solution until it is discharged, when the blood returns again to the lungs, or else escapes through the skin.

Thus it appears that respiration is a process of combustion, in which the fuel is sugar, and the smoke carbonic acid and aqueous vapor. I need not dwell on a fact so universally known as the presence of carbonic acid in the breath. All, however, may not know how large is the volume of this gas which they daily exhale. It varies with age, sex, food, health, and a variety of other circumstances. In a full-grown man the weight of carbonic acid evolved from the lungs varies from one to three pounds in twenty-four hours, which is equivalent to from nine to twenty-seven cubic feet. During the present Lecture the amount of carbonic acid which has been exhaled into this room by my audience is equal to at least seven hundred and fifty cubic feet,* and would fill a room measuring about nine feet in each direction. From the quantity of carbonic-acid gas exhaled we can very readily calculate the amount of charcoal burnt, which in a full-grown man will vary from five to fifteen ounces in twenty-four hours. Hence, the amount of charcoal which in the form of sugar has been burnt up in the lungs of my audience during the last hour, is equal to at least twenty-six pounds,* which I have had weighed out and placed on the Lecture table, in order to give you an idea of the quantity. Moreover, it has been proved that the quantity of heat evolved by a given amount of charcoal in burning is absolutely the same, whether the combustion be rapid or slow, so that the same amount of heat has been generated in our bodies during the last hour by the slow process of respiration as would have been set free by burning this basketful of charcoal. It is no wonder, then, that the temperature of the body is always so much above that of the air, and that even in the coldest climate the heat of the blood is maintained as high as ninety-six degrees. In regulating the temperature of his body, man follows instinctively the same rules of common sense which he applies in warming his dwellings. In proportion as the climate is cold, he supplies the loss of heat by burning more fuel in his lungs, and hence the statements of arctic voyagers, who have told us that twelve pounds of tallow-candles make

* Calculated for one thousand persons.

only an average meal for an Esquimaux, are not inconsistent with the deductions of science.

Respiration, then, like decay, is a process of slow combustion, in which the oxygen of the air attacks and consumes, even at the ordinary temperature, the sugar in the blood. Let us now compare with it the rapid combustion of the same substance. During this Lecture every robust man present has on an average burnt up the equivalent of about two ounces of sugar. This combustion has taken place so quietly, and has set free the requisite amount of heat so gradually, that we have not been conscious of it. In the blood, where it has been going on, sugar and oxygen, as we have seen, are in close contact. In this crucible I have mixed together just two ounces of sugar and two ounces and a quarter of solidified oxygen, solidified by the force of chemical affinity and bound up in a white salt called chlorate of potash. The oxygen and sugar are therefore here lying side by side, as in the blood, but the conditions of slow combustion which exist in the body not being present in the crucible, they will remain in contact indefinitely, until some external agency is applied. The oxygen is now in its passive condition, but a single drop of sulphuric acid will arouse its dormant energies, and you have instantly one of the most dazzling displays of combustive energy which can be produced by art. The only difference between this brilliant deflagration and the combustion which, during the last hour, has taken place in each of our bodies, is simply this; the heat which in the blood has been imperceptibly evolved during an hour, was here concentrated into a few moments, and therefore produced phenomena of intense ignition. All the other conditions, the material burnt, the quantity of material employed, the products generated, and the amount of heat evolved, are in both cases essentially the same.

On comparing these two phenomena together, reflect for a moment on the false estimate which we are apt to make of the phenomena of nature. The splendid displays of combustion arrest our attention by their very brilliancy, while we overlook the silent yet ceaseless processes of respiration and decay, before which, in importance and magnitude, the grandest conflagrations sink into insignificance. These are but the spasmodic efforts of nature; those, the appointed means by which the harmony and order of creation are preserved. Those of us who have merely studied the brilliant phenomena of nature appreciate but imperfectly the grandeur of its forces, and "those of us who limit our appreciation of the powers of oxygen to the energies displayed by this element in its fully active state, form but a very inadequate idea of the aggregate results accomplished by it in the economy of the world." Contemplate the amount of oxygen employed in the function of respiration alone. Faraday has roughly estimated that the amount of oxygen required daily to supply the lungs of the human race is at least one thousand millions of pounds; that required for the respiration of the lower animals is at least twice as much as this, while the always active processes of decay require certainly no less than four thousand millions of pounds more, making a total aggregate of seven thousand millions of pounds required to carry on these processes of nature alone. Compared with this, the one thousand millions of pounds which, as Faraday estimates, are sufficient to sustain all the artificial fires lighted by man, from the camp-fire of the savage to the roaring blaze of the blast-furnace or the raging flames of a grand conflagration, seem small indeed.

*Amount of Oxygen required Daily.**

Whole population,		1,000,000,000
Animals,		2,000,000,000
Combustion and Fermentation,		1,000,000,000
Decay and other processes,		4,000,000,000
Oxygen required daily,	=	8,000,000,000 lbs.

Tons.

3,571,428 in a day.

1,304,642,357 in a year.

Whole quantity, 1,178,158,000,000,000.

How utterly inconceivable are these numbers, which measure the magnitude of nature's processes, —eight thousand millions of pounds of oxygen consumed in a single day! When reduced to tons,

* Taken from Faraday's Lectures on the Non-Metallic Elements, but correcting an obvious error in reducing the pounds to tons.

the numbers are equally beyond our grasp, for it corresponds to no less than 3,571,428 tons. If such be the daily requisition of this gas, will not the oxygen of the atmosphere be in time exhausted? It is not difficult to calculate approximately the whole amount of oxygen in the atmosphere. It is equal to about 1,178,158 thousand millions of tons; a supply which, at the present rate of consumption, would last about nine hundred thousand years. We need not, therefore, fear that the amount of oxygen in the atmosphere will be sensibly diminished in our day or generation; but then this period, immense as it is, is not to be compared with the ages of geological time. The time which has elapsed since the coal we are now burning was deposited in its beds is to be counted by many millions of years, so that since the coal epoch the oxygen of the atmosphere must have been all consumed again and again. Why, then, has it not all been removed from the atmosphere? Simply because, in the beautiful balance of creation, there is always some recuperative process for every such loss. In the case before us, it is, as we have seen, the vegetation. As fast as our breath, our fires, and the process of decay around us are removing the life-giving oxygen, just so fast it is restored by every green leaf which waves in the sunshine, and by every blade of grass which sprouts under our feet. What the animal removes, the plant restores.

I have before stated that, in the process of decay, the oxygen of the atmosphere, which is active in producing the change, is undoubtedly in a peculiarly modified condition, a condition in which its affinities are highly exalted even at the ordinary temperature of the air; and I also stated that this active condition of the element is apparently maintained by the process of decay itself. This subject has been greatly elucidated by recent discoveries. Of all the known processes of slow combustion, the simplest and the most active is the slow combustion of phosphorus. This familiar substance, used to tip the ends of lucifer matches, if exposed to the moist air, slowly combines with oxygen, shining at the same time in the dark with a peculiar phosphorescent light, whence the name of the substance, from two Greek words, signifying light-bearer. The process is therefore entirely analogous to decay and respiration; but since phosphorus is a chemical element, the change is far simpler, and can be more readily studied, and for this reason it may serve to elucidate those more complex processes of nature.

Ozone.

A few years since, Professor Schönbein, a distinguished Swiss chemist, discovered that, while a stick of phosphorus was slowly burning in a jar of moist air, a portion of the oxygen present underwent a most remarkable change. Without entering into the details of these experiments, I will simply state that, when thus modified, ordinary oxygen seems entirely transformed. The great mass of the oxygen of the air, as you will remember, is wholly devoid of odor, and without action on the most delicate organic structures or the most fleeting vegetable colors; but when thus treated it acquires a very strong and pungent odor, rapidly rusts polished metals, excites decay in organized tissues, and at once bleaches the most permanent dyes. Could there be a more complete inversion of properties? One of the most striking characteristics of this new modification of oxygen is its peculiar odor, and hence Schönbein calls it ozone, from a Greek verb signifying *to smell.* Unfortunately, up to this time we have not been able to prepare ozone except in a very dilute condition, for the most active process with which we are acquainted does not modify more than $\frac{1}{300}$ of the whole volume of oxygen employed; but judging from its action in this diluted state, pure ozone is the most corrosive element known.

It frequently happens that a great discovery supplies the wanting links between a number of obscure facts, and thus adds quite as much to our knowledge by its indirect bearings as by the positive additions it makes to the general stock. So it has been with the discovery of ozone. Every one who has used an electrical machine must have noticed the peculiar smell which follows the electrical discharge. This was formerly supposed to be the odor of the electrical fluid itself; but as soon as ozone was discovered, the odor was recognized at once as belonging to this new agent, and it was soon ascertained that electricity is one of the most efficient means of modifying the oxygen of the air. Moreover, the recent investigations of Schönbein appear to prove that ozone is merely electrified oxygen. He has shown that there are two varieties of ozone, corresponding to the positive and negative electricities, which seem to have the power of entering into chemical union, for when brought together they neutralize each other, and ordinary passive oxygen is the result.

Returning now to the fact that the slow combustion of phosphorus throws a portion of the surrounding oxygen into a peculiar condition, in which it is highly active in producing decay and other processes of oxidation, — it certainly seems probable that decay and respiration, which are also examples of slow combustion, may act on the air in the same way. Moreover, the inference that ozone is the active agent in these processes is also supported by the fact that it is always present, to a greater or less extent, in the atmosphere, although, at most, in exceedingly minute quantities. Observations on the subject have now been made for several years, and from these it appears that the quantity of ozone in the atmosphere varies with the season and with the state of the weather. The amount has been found to be greater at night than during the day, and greatest just before sunrise. It is greater in winter than in summer, and at all seasons it is very greatly increased during stormy weather, especially during a snow-storm. The amount is unusually large during one of our easterly storms, and during a thunder-shower the whole surrounding atmosphere frequently becomes pervaded with its peculiar odor, especially in the neighborhood of the place where lightning has struck.

Ozone, being so highly corrosive, cannot be present in the atmosphere in such perceptible quantities without producing important effects, and some persons have thought not only to refer to it the various processes of slow combustion, but also to trace a connection between the prevalence of various contagious diseases and the excess or deficiency of ozone in the air of the infected district; but these speculations are not as yet based on sufficient evidence, and are not worthy of serious attention.

Without, however, introducing any theories not yet fully established into the line of our argument, this much is clear. Oxygen gas appears in nature in three conditions, or under three manifestations; — first, entirely passive, as in the great mass of the air; secondly, partially active, in the processes of decay and respiration; thirdly, highly active, in the phenomena of combustion. In each of these conditions its properties have been adjusted with infinite skill and delicacy, on the one hand to the thermal and electrical conditions of the globe, and on the other, to the constitution of man and of all organic nature.

Here I must conclude my brief sketch of this wonderful element. If I have succeeded in impressing on your minds some of its more characteristic qualities, if, above all, you have become aware how exactly and delicately these qualities have been adjusted to the scheme of creation, and if you have seen how the smallest permanent change would disturb the result,—this is all that I could hope. It might be expected that the element with which creative power built up the greater part of our globe, leaving only a small excess to constitute its atmosphere, would furnish abundant evidence of design, and how fully is this expectation realized! Would that I might present to you the evidence more forcibly! But it is possible in a popular lecture only to touch at some of the more striking points, and I have felt all the time like a school-boy at play, in spring, in some garden rich in flowers, snatching here and there a few of the more gaudy tulips, which had fully bloomed, but leaving the beautiful and delicate buds all unnoticed. But then these buds of knowledge shall blossom, and when the summer comes, the rose and the myrtle will fill the air with fragrance, and bear a still sweeter testimony of goodness and of love.

LECTURE V.

TESTIMONY OF WATER.

THE atmosphere, as you will remember, consists mainly of two permanent and elementary gases; and having in our last Lecture discussed the functions of its active element, oxygen, it would seem natural to consider next the offices of nitrogen, that most singularly inert gas, which constitutes no less than four fifths of its whole mass; but we shall understand more clearly the complicated relations of this truly wonderful substance, associated as it is with all the higher forms of corporeal vitality, after we are acquainted with two of the remarkable cycles in nature, in which the water and carbonic acid of the atmosphere play a conspicuous part. It is true that these two substances are very variable constituents, and make up at best only an exceedingly small fraction of the whole mass of the air; but nevertheless, they discharge functions no less important than those of oxygen and nitrogen, and we shall find that they are equally rich in illustrations of the wisdom and power of God.

I have already alluded to the fact, that the most striking illustrations of creative wisdom have been discovered in those substances which are the most abundantly distributed through nature, and which are the most intimately associated with man, and this not only because they have been more carefully studied, but also because they have the most important functions to fulfil, and have been especially adapted to their appointed ends. Of no substance is this principle more remarkably true than it is of water. As you well know, water is the liquid of the globe, and, if we except certain transient products of volcanic action, it is the only liquid which exists naturally on its surface. Moreover, it is in constant circulation, and, like the blood in our bodies, is the medium through which nourishment is conveyed to all parts of organized nature, and its life sustained. We should naturally expect that a substance filling so important a place in the scheme of creation would furnish undoubted evidences of design, and it will be my object in the present Lecture to illustrate a few of the more striking examples of adaptation which its qualities present, beginning with the aeriform condition of water as it exists in the atmosphere.

The aqueous circulation.

The condition of the atmosphere of aqueous vapor, which surrounds the globe, differs essentially from that of the permanent gases which are simultaneously present. Oxygen and nitrogen cannot be reduced to liquids even by the intense cold at the poles. It is very different with aqueous vapor. The slightest reduction of temperature, when the air is saturated with moisture, is sufficient to condense a portion of the vapor to water, and to shower it on the earth in drops of rain. On the other hand, when the temperature rises, the heat converts more water into vapor, and the aqueous atmosphere is replenished. Thus it is that the atmosphere of aqueous vapor on the earth is liable to very great fluctuations, from which the Creator has protected the great mass of the air by endowing oxygen and nitrogen with the power of retaining the aeriform condition under all circumstances; and we shall find that the fluctuation in the one case is as important as the stability in the other.

I stated in the last Lecture that our atmosphere may be regarded as made up of three partial atmospheres, simultaneously surrounding the globe, and as was the case with the atmosphere of oxygen, we shall best understand the fluctuations of the aqueous atmosphere if we begin by eliminating for a moment from our thoughts the other two. In order to make the subject clear, it will be necessary for me to dwell very briefly on a few well established facts in meteorology, which, although not very interesting in themselves, will unfold to us some of the beautiful provisions of nature by which the aqueous circulation on the globe is maintained.

If there were no free oxygen or nitrogen gas, the earth would still be surrounded with an atmosphere of aqueous vapor, and we are able to foresee, in some small measure, what the conditions of such an atmosphere would be. Its density at the sea level would depend chiefly on the temperature, and would therefore vary very rapidly with the latitude, and would be constantly changing at the same locality with the alternations of the climate. We are able to determine approximately what the density would be at any given temperature, and a few of the results are included in the following table.

Weight at the Sea Level of one Cubic Foot of Vapor.

Temperature. Fahrenheit.	Weight. Grains.	Temperature. Fahrenheit.	Weight. Grains.	Temperature. Fahrenheit.	Weight. Grains.
0°	0.78	40°	3.09	70°	8.00
10°	1.11	50°	4.28	80°	10.81
20°	1.58	60°	5.87	90°	14.50
30°	2.21				

It is evident from these numbers, that a very small change of temperature would cause immense fluctuation in such an atmosphere. At 0° one cubic foot of the aqueous atmosphere could contain only about three fourths of a grain of vapor, while at 80° it could contain fifteen times as much, and hence, although under the tropics the density of our assumed atmosphere would be comparatively large, there would be almost a complete vacuum at the poles. Into this vacuum the vapor would flow from the equator, and thus in either hemisphere there would result a perfect torrent of vapor rushing towards the north or south. But it is also evident that, as this current became chilled in passing through the cooler climate of the temperate zone the vapor would gradually condense to water, which, falling on the land or on the ocean, would return in time to the equator, ready to begin again the same succession of ceaseless changes.

Although the presence of the air materially modifies, it does not essentially change, the aqueous circulation. The air retards the formation of vapor, but does not prevent it, and at any given temperature the same amount of water will evaporate into a given space, whether it be a perfect vacuum or filled with air. Thus, for example, when air at 80° is saturated with moisture, it contains, as before, exactly 10.81 grains of vapor, and the table just given applies equally well to the actual condition of the globe, covered with its dense atmosphere of oxygen and nitrogen, as to the case just assumed. There is, however, a most important difference between the two conditions,—a difference on which the adaptation of the system of aqueous circulation in the order of nature entirely rests.

Were there no air on the globe, the quantity of vapor would adjust itself almost instantaneously to any variation of temperature, and the maximum amount possible would always be present at any given place. An elevation of temperature would be attended by rapid evaporation, and the amount of water required to fill the space would suddenly flash into vapor; while, on the other hand, a corresponding depression of temperature would be accompanied with an equally sudden precipitation of the excess of water which the air could no longer contain, not in genial showers or diffusive rain, but in terrific torrents, of which the deluging showers of the tropics can give us only a feeble conception; for the drops, falling without resistance, would be as destructive in their effects as volleys of leaden shot.

The rain.

In the actual condition of the atmosphere, the presence of a dense medium very greatly retards these changes, and although it does not alter their essential nature, it moderates their action and mitigates the violence of their effects. An elevation of temperature is followed by an evaporation of water into the air; but the process is comparatively slow, and it is a long time before the air is fully saturated. So also when the air is saturated, a depression of temperature is followed by the condensation of a portion of the vapor into rain; but here again the mass of the atmosphere tempers the abruptness of the transition, and allays its violence. The vapor condenses first into vesicles, which are minute hollow bubbles of water, so light that they float in the atmosphere, and when these bubbles break, and the drops of rain fall, they again are wafted down so slowly through the resisting medium, and alight so softly, that the "soft falling snow and the diffusive rain" have become fit emblems of the beneficence of God, as they give the strongest evidences of his wisdom and skill. Moreover, the glorious clouds, which add so much to the beauty of the landscape, and typify in their virgin whiteness the purity of heaven, are only collections of these vesicles of water floating in the upper atmosphere, and mark the stage of transition between vapor and rain; and, further still, it is probable, as I stated in a previous

Lecture, that it is these same minute vesicles which tint the morning and evening sky with their gorgeous hues and cover our earthly dwelling-place with its canopy of blue.

Again, the presence of the air very greatly retards the aqueous circulation above described, without altering its essential character. There is now the same great difference between the density of the atmosphere of vapor at different latitudes, as if it were the only atmosphere on the globe, and the dense vapor of the tropics tends constantly to flow towards either pole; but as it cannot move without carrying with it the whole mass of the atmosphere, this tendency merely increases the velocity of those great aerial currents already described in a previous Lecture. Still the general fact remains the same. From the whole surface of the globe water is constantly evaporating into the aqueous atmosphere which surrounds it. The heated air from the tropics, heavily charged with moisture, is continually moving towards the colder regions, both of the north and of the south; and as the current thus becomes chilled, the vapor is slowly condensed, and the water showered down in fertilizing rains on the land. Thus it is that those beautiful provisions which we see in the rain all depend on the presence of the air, and result from a careful adjustment of the properties of aqueous vapor to the exact density of our atmosphere. "Hath the rain a Father?" Science, by discovering these evidences of skilful adaptation, has most conclusively answered this question, and the answer is the same now as in the days of Job. "Behold, "God is great. . . . . He maketh small the drops "of water: they pour down rain according to the "vapor thereof."

But what becomes of the rain? Would that I could answer this question satisfactorily. We all understand the general theory of the aqueous circulation, but the deepest philosophy and the keenest science are not able to fathom its details, or to comprehend in their fulness the world of wonderful adaptations which the question unfolds. We all know that the drops of rain percolate through the soil, and collect in natural reservoirs formed between the layers of rock, and that these reservoirs supply the springs. The rills from numerous adjacent springs unite to form a brook, which increases as it flows, until it finally becomes the majestic river, rolling silently on its course. Every drop of that water has been an incessant wanderer since the dawn of creation, and it will soon be merged again in the vast ocean, only to begin anew its familiar journey. If you would gain an idea of the magnitude and extent of this wonderful circulation, you must bring together in imagination all the rivers of the world, the Amazon and the Orinoco, the Nile and the Ganges, the Mississippi and the St. Lawrence, and, adding to these the ten thousands of lesser streams, endeavor to form a conception of the incalculable amount of water which during twenty-four hours they pour into the vast basin of the world, and then remember that during the same period at least four times as much water must have been raised in vapor, and scattered in rain over the surface of the land. Would you form an idea of the importance of this circulation, you must not limit your appreciation to its economical value, as a great source of power, working the mills and the forges of civilized man, and building up vast marts of manufacturing industry, nor must you regard alone its commercial value, bearing as it does on its bosom to the ocean the freights of empires. These applications of power, however important in themselves, are insignificant in extent compared with those mighty agencies which the aqueous circulation is constantly exerting in nature. It has been the great agent of geological changes: here washing away continents, and there building them up; here gullying out valleys, and there smoothing away inequalities of surface; here dissolving out the particles of metals from the solid rocks, and there collecting them together in beds of useful ores. It has cov ered the earth with verdure and animal life, by conveying nourishment to the plant and food to the animal. It sustains our own bodies, for it is a portion of this very circulation which ebbs and flows in our veins, and whose pulsations beat out the moments of our lives; and could I bring together in one picture the infinite number of beneficent ends which it has been made by Providence to subserve, I am sure that you would agree with me that there is not in nature stronger evidence of design than in the adaptations of this simple and familiar liquid.

In order that we may appreciate in some humble measure the force of this evidence, let us consider some of the qualities of water; but at the same time let us not forget that the strength of our argument lies not so much in the fact that each property has been skilfully adjusted to some specific end, as it does in the harmonious working of all the separate details. Had man creative power, the first would fall within the range of his intelligence; but to adapt the same substance to a thousand different ends, and to adjust each of its properties to a thousand different conditions, covering with their complex network all the known universe, implies a power nothing less than infinite, and an intelligence nothing lower than divine. It is evident, however, that we can gain a knowledge of the general plan only by studying the details, and unfortunately it is to these details that our accurate knowledge is almost entirely confined. We can see, for example, that each property of water has been designed for some specific purpose. We can also recognize the evident fact, that all the properties work harmoniously together in the general scheme of nature; but, in the present state of knowledge, to trace the intimate relations of these properties is frequently as impossible as it is to form a clear conception of the coexistence and harmonious action of all. Yet in these very facts lies the whole force of the argument from design, and it is only the limitations of our knowledge and faculties which weaken the impression on our minds. But were these limitations removed, all argument would become unnecessary, for then reasoning would be exchanged for vision, and in the refulgence of the Divine Presence we should know even as we are known.

Water a constituent of organized beings.

It is a familiar fact, that moisture is an essential condition of organic life; but few persons, I suspect, are aware that this familiar liquid constitutes the greater part of all organized beings. The physical man has been described by one writer as consisting of merely a few pounds of solid matter distributed through six pailfuls of water, and it is a fact that no less than four fifths of these bodies of ours are made up of water. Yet this is a small proportion compared with the amount which enters into the structure of most of the lower animals. Some of these, such as the medusæ,—*sunfishes*,—are little else than organized water. Professor Agassiz obtained from one of the large sunfishes found on our coast, weighing thirty pounds, only two hundred and forty grains of solid matter; and we may safely say that at least nine hundred and ninety-nine parts in a thousand of these singular animals consist of water. Water constitutes to almost as great an extent most of the vegetable products which are articles of food, as will be seen by the accompanying table.

Plums	contain	75	per	cent	of water.
Potatoes	"	75	"	"	"
Apples	"	80	"	"	"
Carrots	"	83	"	"	"
Turnips	"	90	"	"	"
Watermelons	"	94	"	"	"
Cucumbers	"	97	"	"	"

It is evident from these facts, that water is the chief material of which all organized structures are formed, and in studying the aqueous circulation we have already become acquainted with the beautiful provisions of nature by which this life-giving liquid is distributed over the earth, and showered down upon the meadow and forest alike. Without water organic life cannot exist, and where from any local causes the supply fails, there we find a barren wilderness; while, on the other hand, the genial influences of the rain will soon make even "the desert blossom as the rose." It is a remarkable fact of physical geography, that the distribution of water by the aqueous circulation is rendered more effective by the peculiar structure of the continents, and the position of the great mountain-chains.

"The mountain-chains," writes Professor Guyot, in his excellent work on "Earth and Man," "are great condensers placed here and there along the continents to rob the winds of their treasures, and to serve as reservoirs for the rain-waters, and to distribute them afterwards as they are needed over the surrounding plains. Their wet and cloudy summits are untiringly occupied with this important work, and from their sides flow numberless torrents and rivers, carrying in all directions wealth and life."

Thus the mountains, whose majestic forms affect so powerfully the human soul, and which have exerted such an influence on the history of the race, are also among the most beneficent means in the Divine Providence by which the earth has been fertilized and rendered a fit abode for man. Moreover, these mountain-chains have been evidently so distributed as to give the greatest efficiency to the aqueous circulation, and to irrigate the continents most effectively with their fertilizing floods. We cannot therefore suppose that even these ridges on the earth's surface, which are the lasting records of ancient geological changes, were fixed by chance, for they also bear traces of His intelligence who seeth the end from the beginning, and every part of whose works is adapted to every other. "Lord, thou hast been our dwelling-place "in all generations. Before the mountains were "brought forth, or ever thou hadst formed the earth "and the world, even from everlasting to everlasting, "thou art God."

But it is not the mountains alone which condense the vapor of the atmosphere; for under certain conditions the level plains act in a similar way, and distil down the precious drops of dew upon field and meadow, distributing it among

The dew.

the plants with discriminating care for the necessities of each. The dew is simply another phase of the great aqueous circulation, and like the rain it is a persuasive witness of the wisdom of the Divine Disposer, who has adjusted its amount to the wants of the vegetable world. Every one has noticed the deposition of moisture on a pitcher of ice-cold water during a summer's day, and in this familiar fact we have at once an example and an illustration of the simple provision by which, during even the long droughts of summer, the plants receive a partial supply of water, sufficient at least to sustain their life until the later rains bring the autumn fruits to maturity, and stimulate a more vigorous growth.

The explanation of the dew upon the pitcher is very simple. The layer of air in contact with its cold mass is rapidly cooled, and when it can no longer hold all the moisture it contains, the excess is deposited in drops on the surface. Exchange now the pitcher for the earth, and you have at once an explanation of the proximate cause of the dew. After sunset the earth, like the pitcher, cools down the layer of atmosphere immediately in contact with it to such a degree that the whole of the vapor can no longer retain its aeriform condition. As a necessary result, a portion is condensed and deposited upon the surface, and this is what we call dew. But it will be asked, What cools the earth so suddenly after the setting of the sun? for this is not so evident as the cause of the coldness of the pitcher. Certainly not, and the question will lead us to a study of those relations in which the adaptations to be discovered in this natural phenomenon are chiefly to be found.

The earth, as I stated in the second Lecture, is moving with immense rapidity through a space, whose temperature is at least 270 degrees below the zero of Fahrenheit's thermometer, and, like a heated cannon-ball hung in the middle of a cold room, it is continually losing heat by radiation. The dense atmosphere with which it is enveloped, acting, as we have seen, like a blanket, protects the earth from the intense cold of space to a certain extent; but still the constant loss of heat is so great, that, were the sun's rays withheld for a few days, the temperature of the surface-land, even in the tropics, would fall as low as it is now at the poles during the long night of the arctic winter. In the day-time the earth receives from the sun more heat than it loses; but when this great thermal source is temporarily withdrawn, the loss of heat continuing as rapidly as before, the surface becomes quickly cooled, and the deposition of dew follows, as just explained; or, if the temperature falls below the freezing point, the dew is changed to frost.

You must all have noticed that the most copious deposition both of dew and frost takes place on clear nights, and that during cloudy weather this supply of moisture is entirely withheld. The reason is obvious. The earth loses heat by radiation, and the clouds, intercepting the rays, reflect them back to the earth. A shed or any other protection spread over the ground acts in the same way, and it is well known that a covering, however slight, is sufficient to protect tender plants from the blight of the early frosts. Can it then be an accident, a mere result of chance, that the dew is deposited most abundantly where it is needed most, and that this supply of moisture fails only when the clouds promise a more copious draught of liquid nourishment from the rain?

There is still another fact presented by the dew, which is equally suggestive. The heavens do not distil down their liquid treasures upon all objects alike, but the dew is deposited much more abundantly on the herbage, the shrubs, and the trees, which need the refreshing moisture, than on fallow land, the sandy plain, or the beaten road; and here again the cause has been discovered. It is evident from the general theory of the subject, that the largest amount of dew will fall on the coldest surface, and it is equally obvious that, other things being equal, those objects will cool most rapidly which have the smallest supply of heat to lose, and which radiate it with the greatest freedom. Now it has been ascertained by experiment, that the facility of radiating heat depends entirely on the nature of the surface, and the surfaces of leaves have such a remarkable power in this respect, that it would seem as if they were specially designed for the purpose. If next you consider how small a quantity of matter the leaves contain compared with their large radiating surfaces, you will see that there are all the conditions present of rapid cooling. When, therefore, under a clear evening sky, the rays of heat are escaping from all objects into the celestial space, the green foliage soon becomes colder than the barren rocks or the inanimate clod, and receives in consequence a greater supply of dew.

It will be remembered, as I stated in the second Lecture of this course, that the points of leaves have the power of silently discharging the thunderbolts of heaven, and that in consequence every tree acts far more efficiently to avert the stroke of this destructive agent than the best constructed lightning-rod. Is not now the force of this evidence of adaptation very greatly enhanced, when we find that the surfaces of these same leaves have been endowed with an equally remarkable power of radiating heat, by which they are insured a daily supply of moisture when they need it most? Could the adaptation of the structure of the leaf to these two entirely distinct physical conditions of the atmosphere be the result of anything but intelligence? Admitting that we can possibly conceive, — as Mr. Darwin and the modern advocates of the development theory would have us believe, — that under the pressure of circumstances a plant may have the power of changing its structure so as to adapt it to the external conditions, still I think no one will be so bold as to maintain that there can be any brute agency in vegetation endowed with such foresight as to have adapted the material and structure of each leaf from the very first to the physical conditions of the globe, and this, moreover, for the purpose of effecting ends so remotely connected with its own organization as the discharge of electricity or the radiation of heat. If this can result from chance, under its modern name of natural selection, then chance is but a counterfeit name of God. Gideon believed that God would save Israel, because the dew fell on the fleece but not on the ground, and afterwards on the ground but not on the fleece, and shall we doubt the reality of the Divine Providence, before whom a similar miracle is repeated every evening, with such beneficent results? If it be the mark of intelligence to be able to fathom and comprehend this wonder of Nature, can it be anything below Infinite Intelligence "who hath begotten the drops of dew"?

I might with advantage enter more into detail in regard to the laws of the distribution both of the rain and of the dew, but time and space forbid. I have been able only to open the subject; yet if I have succeeded in impressing you with the extent of the field which these beautiful phenomena present to your inquiry, it is all that I could expect. We have seen that it is through these familiar channels that liquid nourishment is conveyed to the organic world, and the reservoirs supplied which feed the great river-system of the globe. But we should form a very imperfect idea of the resources of nature were we to limit our regards of the aqueous circulation to this important use. The life-blood of our bodies, which conveys to each muscle the nourishment it requires, when it returns again through the veins, this errand well done, is no less usefully employed in carrying away the portions of the tissues which have been worn out in the processes of life; and where from any cause this last function is not faithfully discharged, and the wasted muscles are allowed to remain in the system, disease and death are the inevitable results. So also is it with the life-liquid of nature, which in the rain and the dew carries food to the whole organic world. When this office has been fulfilled, it returns again to the ocean, washing away those waste products of organic life, which, if they remained, would cause pestilence and general decay. It is true that we cannot trace all the details of this cleansing process; but you need not the aid of science to assure you of the general facts. Let the free flowing of the rain-water be interrupted, and you well know that stagnant pools, breeding pestilence, or deadly swamps, exhaling malaria, are the immediate results. I cannot overstate the importance of this function of the aqueous circulation, or too strongly insist on the evidence of wisdom which the adaptation of the properties of water to this beneficent end implies. It is the great cleansing agent of the world. Wherever it flows, there it purifies, and its limpid streams, clear as crystal, are fit emblems of the purity of heaven. Hence the significance of this liquid in all religious systems. The ancient Egyptians worshipped the water of the Nile, and the Hindoo idolaters of the present day rever-

Water the great purifier.

ence with equal devotion the water of the Ganges. Passing to Judaism, we find the washing with water enjoined as a sacred duty by the Hebrew law; and lastly, in the Christian dispensation the pure liquid has become the medium of its most sacred rite, and the outward washing of baptism typifies that inward "washing of regeneration," by which alone man is saved.

Glancing now for a single moment at the æsthetic aspects of the subject, consider what sources of pleasure the varied phases of the aqueous circulation furnish, and what an influence on the soul of man they are calculated to exert. The bubbling spring, the purling rill, the murmuring brook, the sparkling cascade, the roaring torrent, the majestically flowing river, are familiar images of poetry, and the occasions of mental emotions which all have experienced and none can fully describe, while the mighty cataract and the ocean-storm are among the sublimest aspects of nature, and inspire the beholder with reverence and awe. When, now, you reflect that the chords of the human soul have been so strung as to vibrate in sympathy with these emotions of the material creation, and that thus the aqueous circulation has been made a means of instructing and elevating the human race, can you refuse to accept the evidence of wisdom and goodness which a system so far-seeing in its design, and so beneficent in its results, affords?

The mechanism of nature differs, as we have seen, from the creations of human ingenuity, in the fertility of its resources. Man combines numerous means in order to produce a single end; but in nature the most varied and apparently incompatible results flow from a single design. In God's works the means are employed, not as we use them in the poverty of our resources, but from the exuberance of riches. To use the language of another, "All the means are ends, and all the ends are means"; and the grand result is an harmonious system, in which every part is a whole, and where the whole that is known is felt to be only a very insignificant part. Such is the character of the aqueous circulation, which we are now studying, and assuredly the numerous results we have already seen flowing from this simple mechanism are sufficient to mark the system as Divine; but we have not as yet exhausted its resources. Indeed, we have been all the time looking at only one side of the design, and there is a whole set of adaptations yet unnoticed, which are no less important in the scheme of organic nature than those we have so briefly considered; and when we have become acquainted with these, we shall find still other phases of this boundless plan presented to our view, and not until man ceases to learn by study, or the waters cease to roll, will the subject be exhausted.

Agency of water in distributing heat.

We have thus far only considered the agency of the aqueous circulation in distributing over the earth the chief constituent of all organic matter, together with some of the secondary ends which the river-system of the globe subserves. But there is another condition of organic life no less essential than moisture. The animal kingdom is absolutely dependent on the vegetable, and plants cannot grow except within a limited range of temperature. Therefore, unless during at least a portion of the year the amount of heat supplied is sufficient to maintain the temperature of the climate within the required limits, organic life cannot exist in that region. Now this familiar substance, water, has been endowed with most remarkable and unusual properties, by which the aqueous circulation has been made a great means of distributing heat, and thus of sustaining organic life in vast tracts of country where otherwise it could not exist; and it is to this class of its adaptations that I wish next to call your attention.

One of the prominent inventions of modern times is the method of heating large buildings by steam.

Latent heat of steam.

You must all have seen the apparatus. There is first the boiler, where the steam is generated by the combustion of fuel; then pipes, by which it is distributed to the different rooms; next the iron radiators, in which the steam is condensed to water, and during this change gives out heat, which is radiated from the corrugated surface of the iron; and, lastly, the return pipes, through which the condensed water flows back to the boiler, ready to start again on the same journey. Every one is familiar with these external aspects of the apparatus; but all may not know that the efficiency of the method depends entirely upon a remarkable quality of water, a quality which is not possessed to the same degree by any other known liquid. Were you to test with a thermometer the temperature of the water in the boiler and that of the steam rising from it, you would be surprised to find,—if you were not forewarned of the fact,—that they were both at precisely the same point; and yet in order to change one pound of boiling water into one pound of steam it is necessary to burn up sufficient coal to raise the temperature of ten pounds of ice-cold water to the boiling point. The coal which is burning under the boiler does not raise the temperature of the water. Press the fire ever so hard, you cannot increase the temperature either of the water or of the steam by a single degree. The effect of increasing the fire will be only to generate steam more rapidly, for the whole of the immense amount of heat set free by the burning fuel is absorbed by the boiling water in changing into steam. But this heat is not lost. It remains latent in the steam, is carried by it into the different rooms, and there, when the steam changes back again into water, it is all given up, without the slightest diminution, diffusing its genial warmth through the house. The steam, therefore, is merely the vehicle by which heat is carried over the building. The heat comes from the burning fuel in the cellar, and originally it came from the sun; for the coals burning under the boiler are merely fagots, as it were, of condensed sunbeams, gathered by the plants of some ancient geological epoch, subsequently fossilized and preserved in the earth for our use

The steam merely acts the part of a common carrier; but what I wish you to notice is the fact that steam is peculiarly fitted for the work, because it has been made capable of holding so large a quantity of heat.

Your attention perhaps has been called to the efficiency and economy of this method of heating, you have admired its neatness and absolute safety from fire, and have been delighted with the softness of the temperature which it diffuses through the rooms. Or, if you have examined more closely the details of the apparatus, you must have been struck with the ingenuity of the adjustments by which it is self-regulated. Yet this is no new invention. A similar apparatus, on a vastly grander scale, working with far greater economy and efficiency, and provided with adjustments of wonderful delicacy, which perfectly regulate its action, and which never fail and never wear out, has been at work ever since the dawn of the creation, and is at this moment softening the inclemency of our Northern winter.

The general aqueous circulation is a great steam-heating apparatus, with its boiler in the tropics and its condensers all over the globe. The sun's rays make the steam, and wherever dew, rain, or snow falls, there the heat, which came originally from the sun, and which has been brought from the tropics concealed in the folds of the vapor, is set free to warm the less favored regions of the earth. This apparatus of nature, although so much simpler, and working without pipes, iron boiler, or radiators, is exactly the same in principle as the steam-heater, which may be seen at work in almost every large factory. It is true that the atmospheric vapor is a much better vehicle of heat than ordinary steam, and it is also true that this thermal application is but one of the hundred uses of the aqueous circulation; but still the general method is the same, and both systems owe their efficiency to the unique property with which water has been endowed. It is true that other liquids in changing into vapor absorb heat, but the heat stored up in these vapors is vastly less than that in steam, and it must be noticed that, of all created forms of matter, this familiar liquid, which fills the ocean, which distils upon us in the rain, and which flows in the rivers, is the only substance which has been thus especially endowed. Is this an accidental concurrence of circumstances? or is it, on the contrary, the work of Infinite Wisdom? We regard, and with reason, the beautiful invention of man, by which our dwellings are warmed, as an evidence of intelligence; and can we refuse to recognize the existence of that higher Intelligence, which not only adjusted the more perfect system of nature, but also created the properties of water, on which the efficiency of both depends?

Having considered that peculiar quality of vapor through which the aqueous circulation becomes an important means of distributing the sun's heat over the surface of the globe, we might next discuss more at length the extent of its influence, and examine in detail the ingenious system of checks and balances by which the action of this great heating apparatus is regulated, and its constant working secured; but here, as before, having glanced at the main points, I must leave it to your study to fill the unavoidable blanks, and pass on to consider another special property of water by which a similar result is secured.

Capacity of water for heat.

The amount of heat required to raise the temperature of a pound of water, or of any other substance, one degree, is capable of exact measurement, and the quantity has been determined experimentally for almost every known substance. These experiments have led to a remarkable result, to which I alluded in a former Lecture. It appears that, when water is heated through a given number of degrees, it absorbs more than twice as much heat as any other substance except hydrogen gas,—which is one of its constituent elements,—and more than ten times as much as iron and most of the metals. It is not probable that many of my audience have verified this striking result, but you all know how long it takes to boil a tea-kettle, even over a brisk fire, and have, therefore, some conception of the amount of heat which cold water is capable of absorbing. This familiar experience shows that water has a very great capacity of holding heat, and accurate experiment has proved, as just stated, that at the same temperature water contains more than twice as much heat as any other solid or liquid known.

The importance of this simple provision will appear, if you reflect that it makes the ponds, the lakes, and the oceans great reservoirs of heat. It not only requires a vast amount of heat to warm one of these large bodies of water, but when once warmed they cool very slowly. Hence the marked difference between the oceanic and the continental climate in the same latitude. During the summer the ocean eagerly absorbs the heat of the sun's rays, which are showered upon it in such profusion; but water has so great a capacity for heat, that the ocean, nevertheless, does not grow very warm, and, moreover, a large amount of the heat it receives is carried away by the vapor which is constantly rising from its surface. In winter, on the other hand, the water gives up its heat to warm the colder air; but it contains such an inexhaustible supply, that the loss does not materially lower its temperature. There results in consequence a great uniformity of temperature, in which the air, by its perpetual contact with the surface of the water, necessarily shares, and this uniformity extends, in a greater or less degree, to the climate of all islands and sea-board districts. It is quite different with the surface of continents. There the soil becomes rapidly heated under the vertical rays of a summer's sun, and, as its particles are immovable, the surface-layer soon rises to a high temperature, while, on the other hand, in winter it is cooled by radiation with equal rapidity; and this is the cause of those extremes of heat and cold which characterize all countries of the temperate zone removed from the influence of the ocean. The oceanic climate is moderate, while the continental climate is excessive. During the day, under the same circumstances, the land is warmer than the sea, and colder during the night, or, taking the different seasons, the land is warmer than the sea in summer and colder in winter. These general principles have been verified by the extensive series of meteorological observations which, during the last twenty-five years, have been made all over the civilized world. You will find an excellent abstract of the results in Professor Guyot's work on "Earth and Man," before referred to. I have time only to cite a few familiar facts in illustration of my subject, which I will give nearly in his words.

"On the coast of Cornwall shrubs as delicate as the laurel or the camellia are green through the whole year, while under the same latitude in the interior of the continents, the most hardy trees can alone brave the rigor of the winter. But on the other hand, the mild climate of England cannot ripen the grape, although almost under the same parallel grow the delicious wines of the Rhine. At Astrachan, on the northern shore of the Caspian, as Humboldt tells us, the grapes and fruits of every kind are as beautiful and luscious as in the Canaries and in Italy; the wines have all the fire of those of the South of Europe, although in the same latitude, at the mouth of the Loire, on the Atlantic sea-coast, the vines hardly flourish at all. But while in the South of France the winter is a perpetual spring, the summers of the Caspian are succeeded by a winter of almost polar severity."

I might multiply illustrations, but these are sufficient to show how the remarkable property of water which we are considering tends to equalize the climate of the globe.

This influence of water is very greatly increased by the oceanic currents, which, like the winds, are set in motion by the heat of the sun, and are constantly carrying the warm waters of the tropics toward the poles. One of the most remarkable of these currents is the Gulf-Stream, which flows near our coast, and which diffuses the warm waters of the Caribbean Sea and the Gulf of Mexico over the Northern Atlantic, depositing on the shores of Scotland and Norway the plants and seeds of the tropics. It is solely the heat which these waters bring with them from the equator that has made the island of Great Britain so great a centre of commerce and civilization; for it must be remembered that the latitude of England is the same as that of Labrador, and, were it not for the influence of this ocean current, her soil would be equally desolate and barren. If the configuration of our Western Continent were only so slightly changed as to give a passage to the great equatorial current through the present Isthmus of Panama, a change insignificant in comparison with those which have heretofore taken place, "the mountains of Wales and Scotland would become again the abode of glaciers, and civilization would

Oceanic currents.

disappear before the invasion of arctic cold."* So also it is to the enormous mass of heated water which the Gulf-Stream pours into the seas surrounding Northern Europe that Sweden and Norway owe their temperate climate, while at the corresponding latitudes on our own continent the land is shrouded in eternal ice and snow.

Point of maximum density.

But, my friends, all these provisions for distributing heat over the earth's surface would have been insufficient to maintain organic life in our Northern climate, were it not for still another remarkable property with which water has been endowed,—a property even more entirely unique than either of those we have studied, and one which seems to be an exception to the general laws of nature. The familiar cycles of organic life, both in animals and plants, are intimately associated with the succession of the seasons, and this in its turn depends on the inclination of the earth's axis to the plane of the ecliptic, and on the great primary laws by which this axis is constantly maintained in a position parallel to itself during the revolution of the planet around the sun. To these fundamental conditions in the formation of the solar system the whole constitution of organic life on the earth has been adjusted; and Professor Whewell, in his excellent Bridgewater Treatise, has discussed at length the evidences of design which this circumstance affords. It would be foreign, however,

* W. Hopkins's Address before the Geological Society of Great Britain.

to my plan, to consider these evidences here; but, assuming the succession of the seasons as a part of the order of creation, and as a means of adapting a larger portion of the earth's surface for the habitation of organized beings, it is evident that the higher forms of organic life could be sustained in these Northern regions only by furnishing to the plants and animals an adequate protection against the intense cold of winter, and thus preserving the growth of one summer until the returning sun awakens new life in the succeeding spring.

The required protection has been provided by making a most marked exception to the general laws of expansion in the case of water. It is the general law of nature, that all substances are expanded by heat and contracted by cold, and water forms no exception to the general rule, except within certain very narrow limits of temperature, shortly to be noticed. Indeed, were it not for the expansion, we could not readily either heat or cool a large mass of liquid matter. All liquids are very poor conductors of heat, and can be heated only by bringing their particles successively in contact with the source of heat. When you set a tea-kettle over a fire, the first effect of the heat is to expand the particles of water resting on the bottom of the kettle, which, being thus rendered specifically lighter, rise, and are succeeded by colder particles, which are heated and rise in their turn; and thus a circulation is established by which all the particles are successively brought in contact with the heated bottom of the kettle, and in the course of time the temperature of the whole mass is raised to the boiling point. The case is similar when you add ice to a pitcher of water in order to cool it. The water at the top of the pitcher, in contact with the ice, is of course cooled, and, being thus rendered specifically heavier than the water below, sinks and gives place to the warmer water, which is cooled and sinks in its turn, and thus as before a circulation is established, which continues until the temperature of the whole water is reduced to 40°. But at this point the circulation is entirely arrested; for, in consequence of its singular constitution, water at 39° is lighter than water at 40°, and consequently remains at the top. And so it is as the temperature sinks toward the freezing point. The colder the water, the lighter it becomes, and the more persistently it remains at the surface. Hence, although the upper layers of water may be readily cooled to the freezing point, yet, in consequence of its poor conducting power, the great body of the liquid below will remain at the temperature of 40°.

The cold atmosphere of winter acts upon the ponds and lakes exactly as the ice on the water in the pitcher. They also are cooled from the surface, and a circulation is established by the constant sinking of the chilled water until the temperature falls to 40°. But at this point, still eight degrees above the freezing point, the circulation stops. The surface-water, as it cools below this temperature, remains at the top, and in the end freezes; but then comes into play still another provision in the properties of water. Most substances are heavier in their solid than in their liquid state, but ice, on the contrary, is lighter than water, and therefore floats on its surface. Moreover, as ice is a very poor conductor of heat, it serves as a protection to the lake, so that at the depth of a few feet, at most, the temperature of the water during winter is never under 40°, although the atmosphere may continue for weeks below zero.

Ice lighter than water.

If water resembled other liquids, and continued to contract with cold to its freezing point, — if this exception had not been made, the whole order of nature would have been reversed. The circulation just described would continue until the whole mass of water in the lake had fallen to the freezing point. The ice would then first form at the bottom, and the congelation would continue until the whole lake had been changed into one mass of solid ice. Upon such a mass the hottest summer would produce but little effect; for the poor conducting power would then prevent its melting, and instead of ponds and lakes we should have large masses of ice, which during the summer would melt on the surface to the depth of only a few feet. It is unnecessary to state that this condition of things would be utterly inconsistent with the existence of aquatic plants or animals, and it would be almost as fatal to organic life everywhere; for not only are all parts of the creation so indissolubly bound together that, if one member suffers, all the other members suffer with it, but moreover the soil itself would, to a certain extent, share in the fate of the ponds. The soil is always more or less saturated with water, and, under existing conditions in our temperate zone, the frost does not penetrate to a sufficient depth to kill the roots and seeds of plants which are buried under it. But were water constituted like other liquids, the soil would remain frozen to the depth of many feet, and the only effect of the summer's heat would be to melt a few inches at the surface. It would be, perhaps, possible to cultivate some hardy annuals in such a climate, but this would be all. Trees and shrubs could not brave the severity of the winter. Thus, then, it appears that the very existence of life in these temperate regions of the earth depends on an apparent exception to a general law of nature, so slight and limited in its extent that it can only be detected by the most refined scientific observation.

Moreover, this exceptional property is united in water with another quality, which greatly aids in preserving vegetable life during the winter months. We shudder at the thought of snow, but nevertheless it affords a most effectual protection to the soil, forming as warm a covering as would the softest wool. Water in all its conditions has been made a very bad conductor of heat, and snow is ranked with wool among the poorest of conductors. Heat, therefore, cannot

Snow a poor conductor of heat.

readily escape from a snow-covered soil, and thus its temperature is prevented from falling materially below the freezing point, however great the severity of the season. Notice now how, when winter sets in and the cold increases to such a degree as to endanger the tender plants, Nature promptly spreads her great frost-blanket over forest, prairie, meadow, and garden alike, so that all may slumber on in safety until the sun returns and melts away the downy covering, when the buds break forth again and the trees put on a new mantle of living green.

This leads me to speak of still another remarkable property of the wonderful liquid we are studying; for nature has provided in the constitution of water a most effectual means of tempering the transition of the seasons, and protecting vegetation against the early frosts of autumn or the first deceptive glow of returning spring. In order to freeze a liquid it is necessary to remove from it a certain quantity of heat called the heat of fusion, and the more of this heat a liquid contains, with the more difficulty of course it freezes, and when once frozen the less readily the solid melts. Now water contains a larger amount of heat of fusion than any other liquid yet examined, and in this respect, therefore, it is also peculiarly constituted. And mark how this property tends to produce the result just noticed. As the weather becomes cooler in autumn, our ponds and lakes gradually give up the stores of heat which they contain, until the temperature of the whole mass of water is

Heat of fusion.

reduced to 40°; then the surface-water cools still further to the freezing point; but before it can become any colder than this the water must freeze, and in freezing it will set free four times as much heat as it has already given out in cooling from the temperature of summer (63°) to the freezing point. It is evident, therefore, that freezing must be a slow process. Moreover, it is also a warming process, and although the temperature of surrounding objects can never be thus raised above the freezing point, nevertheless, the immense amount of heat evolved greatly tends to retard the approach of severe cold, and prepares the way for the inclemency of winter. So also, when Spring comes, vegetation is not awakened by her first blushes to be exposed to the blights of the early frosts, and before the snow covering can be melted off the danger is mostly passed. Again, when we consider what devastating floods would sweep the earth were the icy bonds of winter suddenly dissolved, we shall discover still further evidence of the wisdom of that Being who has so adjusted the properties of water that both frost and freshet are the exception, not the rule.

I have said that water presents the only well-established exception to the laws of expansion by heat, and some writers on natural theology have dwelt on this point as one of great importance to their argument. But I cannot think they are wise; for, to say the least, they rest their argument on our ignorance and not on our knowledge. It is true that in the present state of science the anomalous expansion of water near the freezing point seems to be an exception to the general laws of nature; but hereafter this very anomaly may appear to be the natural result of a more general law not yet discovered, or, like the perturbations in the orbits of the planets, may prove to be the strongest confirmation of the very law it now seems to invalidate. Moreover, I do not share in that indefinite dread of natural laws which troubles so many religious minds. To me the laws of nature afford the strongest evidences of the existence of a God, and in their uniformity I see merely the constant action of an omnipresent Creator, who acts with perfect regularity because he acts consistently and with infinite wisdom. I believe that all parts of nature are correlated by laws, and that the wider our knowledge becomes, the more universal these laws will appear. I do not, therefore, regard the constitution of water as something apart from law, and as the evidence of a power coming down as it were upon law to make an exception to it. This is making altogether too much of law. God is not bound by law. He acts wisely, beneficently, and with a definite plan, and the most we can claim for natural laws is, that they are our imperfect human expressions of this Divine plan. Moreover, that is a far nobler view of God's wisdom which supposes him to be able to harmonize special adaptations with general laws. What I find so remarkable in the constitution of water is, not that it is an exception to the general laws of nature, but that, while filling its place in the general plan, it has been endowed with such extreme properties, and that in each case the peculiar property has special adaptations at once so complex and so important. Not only has water this exceptional property of expanding when other liquids contract, but moreover, of all known substances, it has the greatest capacity for heat; so also, when changing into vapor, it absorbs more heat than any other liquid; again, it is far lighter in the solid than in the liquid state; and lastly, it contains the largest amount of heat of fusion as yet observed in any substance. All this may be in harmony with general laws. I have no doubt that it is; but the existence of the law does not in the least impair the significance of the fact, that in each of these respects water has been peculiarly constituted. This one liquid of the globe, which covers more than three fourths of its surface, which circulates through all its channels, which percolates through all its pores, which constitutes three fourths of all organized beings, has been endowed with these four pre-eminent qualities, on each of which the whole order of terrestrial nature may be said to depend. I cannot conceive of stronger evidence of design than this; and if these facts do not prove the existence of an intelligent Creator, then all nature is a deception and our own faculties a lie. Yet, my friends, this is only a small part of the evidence of design which science has discovered in this familiar liquid. I might occupy several Lectures with this subject alone, but I have time only to glance at two more striking facts.

Water is the most universal solvent known, and there are but few substances which are not to a greater or less degree dissolved by it. Those which we call insoluble generally differ from the rest only in degree. Thus, all lime rocks dissolve to a limited extent in spring water, and the same is also true of almost all mineral substances. The magnificent crystals which we frequently find in the rocks are formed in almost every case by a deposition of the mineral substance from a state of solution in water. The feeble solvent power of the water for these substances is made up by the large volume of the solution, and the length of time occupied in the process of crystallization. Many of the large crystals which may be seen in cabinets of minerals have been unquestionably thousands of years in formation. And not only does the solvent power of water stud the cavities of the rocks with gems, but it is also constantly producing most important changes in the rocky structures of the globe itself, here cementing together the loose sands, and there converting the soft clays into firm and solid rock.

Solvent power of water.

Again, the solvent power of water extends to aeriform as well as to solid substances, so that the gases composing the air pervade the lakes and the oceans as well as the atmosphere. Indeed, it is on the gases dissolved in the water that all the aquatic plants and animals live, and the members of the various finny tribes breathe the free oxygen dissolved in the water, as we breathe the oxygen of the air. Again, the process of respiration is essentially the same with these lower animals that it is with us, and the structure of their organs has been adjusted to the amount of this life-sustaining element, which water is capable of dissolving. Moreover, the power which water possesses of dissolving oxygen is much greater than its power of dissolving nitrogen, and hence the air dissolved in the ocean is proportionally much richer in oxygen than our atmosphere. This is undoubtedly another quality with which water has been endowed in order to render the oceans, the lakes, and the rivers a fit habitation for that world of organic life which modern zoölogy has revealed. That we are unable to trace all its relations, is evidently owing to the imperfection of our knowledge. But here a new field of study opens before us, which, when fully explored, will undoubtedly prove as rich in the evidence of design as the atmosphere itself.

It is not, however, merely as a solvent, that water is an important agent in the great labora-

Water as a chemical agent.

tory of the world. I have already stated to what an extent all animal and vegetable substances are composed of water, and that some, such as the jelly-fishes among animals, and the gourd family among plants, may be said to be living forms of water. But we should entertain a very erroneous conception of the condition of the water in these animal and vegetable structures, were we to regard it as so much dead material, building up the form like the bricks in an edifice. This water is in constant circulation, conveying nourishment to all the parts, and at the same time removing from the system those tissues which have fulfilled their functions and become effete. It is being constantly decomposed, and as rapidly again reformed, assuming the most protean conditions, and administering to the functions of the animal economy in a thousand ways.

As a constituent of inorganic matter, water is no less important than it is in organized being. A substance so bland as water, and apparently so entirely inactive, which fills the most delicate vegetable cells, and penetrates the finest capillaries of the body,—whose minuteness and delicacy no art can approach, nor imagination scarcely conceive,—yet without affecting either in the slightest degree, we should suppose would be endowed with no affinities, and capable of exerting no chemical force. Yet what is the fact? In attempting to classify chemical compounds, I have studied with care the chemical history of water, and its relations to other substances, and it is still to me a perfect enigma in nature. For, so far from being that inert material which its bland exterior would seem to indicate, it is among the most important of chemical agents, forming some of the most stable compounds, and surprising the chemist by the strength of its affinities. Not only is water a common constituent of most crystalline salts, and an essential ingredient of many of the powerful acids which are used both in the arts and in the chemist's laboratory, but it makes also a not unimportant part of the rocky crust of the globe. Besides forming the immense deposits of ice which perpetually surround either pole, and the glaciers which creep down the high mountain slopes, we find that water enters as an essential ingredient into the composition of talcose and chlorite slate, gypsum, serpentine, soapstone, and other rocks. Moreover, water is the medium in which most chemical processes take place, and throughout all geological history it has been producing the most fundamental changes in the composition of the earth's crust, the extent of which geologists are only of late beginning to appreciate. It is now supposed that granite and similar rocks, which were formerly regarded as products of igneous fusion, have been really formed from loose beds of mud and clay, through the transforming power of this wonderful and powerful agent.

When, lastly, we consider the composition of water, our wonder is still further increased; for it consists of two permanent gases, condensed by the force of chemical affinity to the liquid condition. With one of these, oxygen, you are already familiar. The other is a light, combustible gas, called hydrogen, fourteen and a half times lighter than air, and by far the lightest form of matter known. One cubic foot of

Composition of water.

water yields more than eighteen hundred cubic feet of a mixture of these two gases, and so persistently do they retain their aeriform condition, that not even a pressure of twenty tons on the square inch is sufficient to reduce them to liquids. Yet, immense as this pressure seems, requiring all the mechanical skill of man to apply it, that force must be still greater which is constantly acting in every drop of water to hold these highly elastic gases in the liquid state. It is difficult to estimate the magnitude of such power, as our only standard of measurement is the quantity of some other force, equally immeasurable, which is required to balance the first. Water is easily decomposed by electricity, and the amount of this agent required to force apart its constituents may perhaps give you some imperfect conception of the magnitude of that power by which they are so securely imprisoned. The statement may seem incredible, but yet it has been proved by Professor Faraday, that it requires more electricity to decompose a drop of water than to charge a thunder-cloud.

What a revelation of power we have here! In every drop of water there is a constant striving of the elements to escape; they are exerting a force to break the bonds that unite them, which can be measured only by the power of concentrated thunderbolts, and yet this immense force is kept in check by a force of equal power, and so exactly balanced that not the slightest disturbance can occur. When, now, I reflect that this chemical force has produced a result which all human strength, aided by the appliances of modern art, has been unable to effect, — that in every particle of water the force is still acting, so that every rain-drop which falls, animates the river, or swells the ocean, is a silent monitor of human weakness, I am overwhelmed by that mystery in nature, which here as elsewhere ever points upward to the Infinite, and thus silently teaches that the mighty influence which binds the atoms of the rain-drop is merely the manifestation of His ceaseless power who holdeth "the waters in the hollow of His hand."

It is a very common mistake to suppose that the grand in nature is to be seen only in its great water-falls and its lofty mountains, and that, in order to appreciate the omnipotence of the Creator, we must cross oceans and explore continents; for, to the intellectual eye, there is more evidence of omnipotence, more real grandeur, in a single rain-drop, than in the roar of Niagara or in the magnitude of Mont Blanc. The more I study the evidence of design in this simple liquid, the more I find there is to learn, and I feel the utter inadequacy of any language to convey the full and complete idea. Review for a moment the examples of adaptation which have been so briefly noticed. Remember that water is the liquid of our globe, and the only liquid which exists in any abundance on its surface. The total amount of all other liquids is in comparison but as "a drop of a bucket." Consider, next, that its specific gravity has been so adjusted that our ships float, and the oceans are made great highways for the nations; that it is easily converted into vapor, and as easily condensed to fertilizing rain and refreshing dew, which nourish the growing plants, fill the springs, and keep the rivers, the great arteries of the globe, in circulation; that at a comparatively low temperature it is changed into highly elastic steam, which, imprisoned by man in his iron boilers, has become the great civilizer of the world; that it has been so exceptionally constituted that the great mass cannot be cooled below forty degrees, and again has been made such a poor conductor of heat that, when the surface is frozen, the very ice becomes a protection against the cold; that to this same liquid there has been given a very great capacity for heat, and that thus it has been made the means of tempering materially the climates of the globe. Add to this, that water has been made an almost universal solvent; that from the substances it holds in solution the crustacea form their shells and the coral polypi build their reefs; that it fills the cavities of the rocks with gems, and their fissures with useful ores. Besides this host of wonderful mechanical adaptations, remember that water has been made a chemical agent of great energy and power; that there have been united in it the apparently incompatible qualities of blandness and great chemical force; that, although in the laboratory of nature it corrodes the most resisting rocks, it also circulates through the leaflets of the rose and the still more delicate human lungs; that it forms the greater part of all organized beings, from the lichen to the oak, and from the polyp to man. Reflect, now, that these are only a few of the grosser qualities and functions of this remarkable compound, gleaned here and there from many others no less wonderful, and you will form still but a very imperfect conception of the amount of design which has been crowded into it. Attempt to find a liquid, which, if in sufficient quantity, might supply its place, and you will be still further impressed by this evidence of intelligence and of thought. Of all the materials of our globe, water bears most conspicuously the stamp of the Great Designer, and as in the Book of Nature it teaches the most impressive lesson of His wisdom and power, so in the Book of Grace it has been made a token of God's eternal covenant with man, and still reflects His never-fading promise from the painted bow.

LECTURE VI.

TESTIMONY OF CARBONIC ACID.

When standing by some quiet mill-stream, have you ever speculated on the origin of the power which is animating the machinery of that great factory on its banks, spinning and weaving the crude cotton into miles of cloth every week? And if, perchance, you have visited Niagara, did the thought strike you, when gazing up at those floods of water, which came tumbling over the rocky cliffs, and plunging into the seething sea at your feet, that similar floods had been pouring over that ancient river-bed for countless ages, yet without diminishing the inexhaustible supply? Or, if it has been at once your privilege and your terror to witness that grandest sight of nature, a violent storm at sea, have you been impressed by the untiring might of that mysterious agent which impels the raging winds and upheaves the roaring billows? Whence can come all the power? and what keeps in motion that wonderful aqueous circulation, which we studied in the last Lecture?

Although the origin of this never-ceasing motion may be still concealed, we have at least traced back the power to a proximate source in the great central luminary of our system. It is the gentle influences of the sunbeam which raise the water in vapor, and it is the same solar rays which keep in motion the great aerial currents, spreading the clouds over the earth, and distilling down their liquid treasures "to satisfy the desolate and waste ground, and to cause the bud of the tender herb to spring forth." Incredible as it may appear, it is actually the sun which weaves the cloth, which feeds the fountains of Niagara, and it is his delicate rays which rule in the tempest and direct the storm. But there are other influences of the sunbeam still more subtile than these, and there are other cycles of changes, as grand as the aqueous circulation, of which the sun is also the ever-active cause.

Circulation of carbon.

Referring to the table before given, representing the composition of the atmosphere, you will notice that the great aerial ocean contains more than five million billions of tons of an aeriform substance called carbonic acid. This amount, although absolutely very great, is nevertheless only a small fraction of the whole atmosphere, making up less than a thousandth of its total mass. A cubic foot of air does not contain more than a quarter of a grain of carbonic acid; yet there is not one of the atmospheric constituents which is more intimately associated with organic life, or which discharges more important functions. Although itself a colorless gas, carbonic acid consists of ordinary black charcoal combined with oxygen, and these elements are united by one of the strongest affinities known in nature. Yet, intense as this force is, the power of the sun is greater, and his rays, acting on the green leaves of the plants, are constantly decomposing the gas and liberating the charcoal, to be incorporated into the various forms of vegetable life. Here, however, it remains only for a brief period; for when the plants have finished their allotted term of life, the charcoal again unites with oxygen, and in the form of carbonic acid is restored to the atmosphere by the process of combustion or decay. But frequently, before these destructive changes complete the cycle, the charcoal undergoes a further transformation, and through the process of digestion becomes a part of the body of the animal. Yet this transmutation, as a general rule, only hastens the final result; since the processes of animal life are very rapid, and sooner or later the charcoal is burnt up in the body and breathed out into the atmosphere, ready to renew the same series of changes. I wish to ask your attention this evening to the evidences of design which may be discovered by studying this wonderful circulation of carbon; and we shall find that the properties both of carbon and of carbonic acid have been most carefully adjusted to the part which they play in the great scheme of nature. We might begin our study at any link of this endless chain of phenomena; but in order to bring the subject into an orderly succession to our previous trains of thought, let us return to the phenomena of combustion, which we considered in the third Lecture, and study the details of this familiar process a little more closely.

All fuel, without exception, contains charcoal, or, as the chemists call it, carbon. Wood, soft coal, oil, wax, and similar combustibles, which burn with flame, contain besides carbon a variable quantity of hydrogen and oxygen; but hard coal, coke, and common wood charcoal are almost pure carbon. The adaptations of each of these classes of combustibles demand special notice, and let us begin by studying the evidences of design which are to be found in an ordinary hard-coal fire; and while, in imagination, we are preparing the fire to be lighted on the grate, we may study with profit some of the external properties of the coal, for even they betray the master hand of the Great Architect.

Composition of fuel.

Examining closely this lump of charcoal, you will notice that it retains all the delicate structure of the wood from which it was prepared. Here is the fibrous bark next the sap-wood, and then the annual rings, all as on a stick of beech-wood; and if you will take the pains to make a thin section of the charcoal, you will find, on examining it with a microscope, that the minutest cells have been preserved. You know how charcoal is made. The wood is exposed to a high temperature in the charcoal mounds or furnace, by which the gases which it contains are driven off, while the charcoal, not being volatile, remains behind. Here, then, is a remarkable fact,—that, although the wood has been exposed to a red heat in the process of carbonization, yet the minutest

Infusibility of carbon.

cells have not been destroyed; and it illustrates an equally remarkable quality of charcoal, on which, as we shall see, its usefulness as fuel very greatly depends. Carbon, in all its forms, is absolutely infusible. It does not even soften at the highest temperatures which can be attained by art, and it is for this reason that the charcoal retains so perfectly the structure of the wood. Were carbon fusible at a red heat, the charcoal would run together to a shapeless mass in the mounds or furnaces in which it is prepared, and did it even soften at this temperature, the forms of these delicate cells could never have been preserved. Viewed in connection with the volatile qualities of the other elements of organized beings, the extreme fixity of carbon in its uncombined condition is worthy of your special attention. The only other essential elements of organic matter are oxygen, hydrogen, and nitrogen; and these three substances are not only gases, but gases which even by the greatest cold or by thousands of atmospheres of pressure cannot be condensed to the liquid condition; yet so strong is the tendency of carbon to remain solid, that it condenses these gases around itself in every organized substance that exists. Carbon is thus the solid substratum of organized matter, the skeleton, as it were, of every organic form. How evidently, then, has the attribute of infusibility been adapted to this important function which carbon has been appointed to subserve!

Examining again this lump of charcoal which

we are using to kindle the fire, mark that it has a black color and is perfectly opaque. These qualities are so evident to the most superficial observation that they are apt to be passed unnoticed, and yet it is these qualities of charcoal which make it so valuable as the basis of printing ink. All my audience may not know that printing ink is a mixture of lamp-black and oil, and that the letters on a printed page are formed by thin layers of black charcoal spread over the white paper; and charcoal is peculiarly well adapted for this use, since, however finely subdivided, it never loses its dead black color and perfect opacity. But this property of charcoal would be useless to the scholar for diffusing knowledge, were it not combined with qualities still more remarkable, and almost unique. Carbon is not acted upon by atmospheric agents, and, moreover, is absolutely insoluble in any liquid, with perhaps the exception of melted iron. The letters of the first Bible ever printed are as black as they were the day they left the types. They have been exposed to the action of atmospheric air for four hundred years, and, were carbon in the slightest degree acted upon by the atmosphere, they would long since have disappeared. As it is, they will endure as long as the paper on which they are printed lasts.

Black color and opacity.

The almost unparalleled insolubility of charcoal is a quality equally important in this relation, for were charcoal even to a slight extent soluble in water, the books of our fathers would have been rendered illegible by the dampness to which all libraries are more or less exposed; and were carbon soluble even in such liquids as alcohol, ether, or the volatile oils, the printed page would not have been, as now, safe from alteration, and all kinds of fraud would have been easy. We justly honor the names of Gutenberg and Faust, whose art has done so much to enlighten and civilize the globe, and we bestow due admiration on those improvements in the art of printing, nowhere more developed than in our own land, which have made the press the great agent of power, and elevated the moral and intellectual above the physical man. But while we pay just tribute to the genius of these benefactors of mankind, let us not forget that greater Benefactor, who was before them all; for in the most familiar qualities of this piece of charcoal, on which the art of printing so greatly depends, there has been displayed since the foundation of the globe an evidence of wisdom and skill before which all human ingenuity sinks into insignificance.

But, my friends, the most remarkable attribute of carbon does not appear in this piece of charcoal; for of all the chemical elements carbon is by far the most protean in its aspects, and charcoal is but one of its many manifestations. In the first place, there are the numerous varieties of coal, including charcoal, coke, lamp-black, and bone-black, all having the same general properties, and most of them partaking more or less of the structure of the organic tissues from which they

Modifications of carbon.

were made. But besides these varieties, which, although differing so much in their outward aspect, have all essentially the same properties, there are two entirely different states of carbon, differing as much from each other and from common charcoal as any two solids possibly could.

Diamond.

Are you aware that the brilliant gem you prize so highly is the same chemical element as these black coals? The diamond is merely crystallized charcoal, and although we cannot crystallize charcoal artificially in our laboratories, and do not even know how the diamonds are made in the great laboratory of nature, yet there is no fact of chemistry better established than this. To those who are not familiar with the results of modern chemistry, it seems almost incredible, and even the chemist can hardly believe the truth while he affirms it. It is at utter variance with the whole doctrine of his science; it cannot be reconciled with any of his previous conceptions, and constantly reminds him of the limitations of his knowledge and the uncertainty of his philosophy. And, turning to the more familiar aspects of the subject, how singular the fact, and how typical of the universality of Christian brotherhood, that He who "hath made of one blood all nations of men," should have also made of the same material the priceless brilliant which adorns the diadem of the prince, and the soot which begrimes the cabin of the humblest peasant! How different the estimation in which these two forms of carbon are held! and yet, if the marks of Divine wisdom can give nobility to a substance, the one is as excellent as the other.

But carbon exists in still a third modification, differing as much from the diamond as that differs from charcoal. Every one who has used a common lead pencil is familiar with graphite, and it is a fact as remarkable as the one just noticed, that the same carbon which forms the letters of a printed page fills also the lines of the pencil sketch. Graphite is simply a modification of charcoal, and if this fact is not so well known as the humble relationship of the diamond, it probably arises from the circumstance that graphite has been generally called plumbago, or black-lead, a misnomer which has given a very erroneous conception of its nature. Graphite is frequently mixed with impurities, but it never contains lead, and in its finest condition it is nearly pure carbon. Compare now graphite with the diamond. Could there be two substances more unlike? the one the softest of minerals, the other the very hardest; the one dull and opaque, the other brilliant and transparent. But besides these external differences they have also a different crystalline form, a different specific gravity, a different capacity for heat, and, in fine, excepting that they are both infusible and combustible, there is not one point of resemblance between them. How then, you will ask, do we know that they are both the same substance as charcoal? Simply because, when

Graphite.

combined with oxygen, they all yield the self-same compound.

All three of the modifications of carbon are combustible, although they take fire at very different temperatures. Charcoal will burn at a red heat, the diamond at a white heat, while graphite requires the highest temperature which can be attained by art. But however different may be the temperatures required, the process is the same in all cases and the result is the same. The burning is simply combination with the oxygen of the air, and the result of that combination is carbonic-acid gas. Moreover, it has been proved by the most careful experiments that a given weight of either substance yields precisely the same weight of carbonic acid. Chemically considered, then, the diamond, graphite, and charcoal are the same substance, although, physically regarded, no substances could be more unlike. Chemical identity, therefore, does not consist in identity of properties, and we must admit that the same chemical element may manifest itself under utterly different physical aspects.

Allotropism.

This remarkable phenomenon, which has been fully recognized only of late years, has been called by chemists *allotropism*,* and the diamond, plumbago, and charcoal are different *allotropic* modifications of the element carbon. Such differences of manifestation, moreover, are not confined to carbon, nor are they exceptional occur-

* Derived from two Greek words signifying difference of condition.

rences among the elements. We have already seen that oxygen may exist in an active and in a passive modification, which stand in as striking an antithesis to each other as the diamond and charcoal, and the same is true of the different conditions of sulphur, phosphorus, and silicon. Again, these phenomena are not limited to the elementary substances, for they have been observed in many compounds as well, and every year enriches our knowledge with fresh examples. In what, then, are such developments to end? If substances so utterly unlike as the diamond, graphite, and charcoal are merely modifications of the same element, why may not all substances be merely different *allotropic* states of a few universal principles, or possibly of only one single essence? Such, and many similar questions, arise in the mind of the chemist while contemplating these obscure phenomena. They cannot be answered in the present state of chemistry, and they throw a degree of uncertainty and doubt on its whole philosophy. I shall have occasion to dwell upon this subject more at length in another Lecture, and have adduced the facts at this time chiefly as further illustrations of that fertility of resources which so strikingly marks all the results of creative skill. To me this characteristic of the works of nature is one of the most convincing evidences of divinity. While studying the simple adaptation of means to ends which we find everywhere around us, we recognize in the plan something analogous to the creations of human skill, and we almost feel a conscious relationship with its Author. But when we consider this incomprehensible power, by which the same substance has been endowed with entirely different and incompatible properties, and not only this, but has been adapted in each condition with equal skill to produce the most opposite and seemingly irreconcilable results, we are also made to feel most keenly, that, although man was created in the image of his Maker, he resembles the Divine Original only as the finite can resemble the Infinite. "For my thoughts are "not your thoughts, neither are your ways my "ways, saith the Lord. For as the heavens are "higher than the earth, so are my ways higher "than your ways, and my thoughts than your "thoughts."

Burning of hard coal.

Of all the properties of coal, the one with which we are most familiar is its combustibility; and while we have been discussing its external properties, the hard-coal fire has been built in the grate, and it is ready to be lighted. The combustion of coal in one or the other of its varieties is the great source of all the artificial heat used by man. Although so entirely passive towards atmospheric agents at the ordinary temperature, yet when heated to a red heat it takes fire and combines with the oxygen of the air with great rapidity. The burning of coal is so familiar to every one that it would seem hardly necessary to dwell upon the subject here. But although the experiment is repeated every day in every grate of this city, and although it has been familiar to you all from infancy, there are, nevertheless, phenomena connected with it which few have observed and still fewer fully appreciated. It is a great mistake, but a mistake too frequently made even by scientific men, to suppose that new knowledge can be gathered only from the unexplored fields of science, when by the most familiar walks of life there are countless riches of truth which the reapers in the hurry of the harvest have passed unnoticed, and which will abundantly reward the careful gleaners. In the coal fire on which you daily gaze, there is enough to be discovered to engross the attention of the most diligent student of nature. Let us see, therefore, if we too cannot learn something new at least to us from the burning coals.

The first fact to which I would call your attention, is the difficulty experienced in lighting coal. In order to kindle the fire we have placed on the bottom of the grate, first, some shavings, then some charcoal, and last of all the hard anthracite coal. We can readily set fire to the shavings with a match, and they in their turn will ignite the charcoal; but it requires the intense heat of the burning charcoal to ignite the anthracite. Charcoal will not burn unless at a full red-heat, and hard coal requires a still higher temperature. But notice now another fact: when once inflamed, the heat evolved by the combination of the carbon with oxygen is sufficient to sustain the temperature at the point of ignition. Here, again, we see

Point of ignition.

most admirably illustrated the adaptation of the properties of the chemical elements to entirely different ends. In order that carbon might serve as the solid substratum of all organized beings, it was necessary that it should be made unalterable by the air within the limits of terrestrial temperature, but at the same time the economy of nature required that it should be made combustible, that is, endowed with strong affinities for oxygen; yet these affinities have been so carefully regulated, that they are called into play only at a high temperature, and are thus placed entirely under the control of man.

Now that the coal is in violent combustion, combining rapidly with oxygen, notice that it burns entirely without flame. We have here rapid chemical combination, with all the phenomena of active burning, and yet no flame, simply because flame is always burning gas, and in a hard-coal fire it is not a gas, but a highly fixed solid, that is burning. Charcoal and anthracite are almost the only combustibles which burn in this way. Most others, even when naturally solids, are converted into gases at a high temperature, and therefore burn with flame; but carbon in all its forms, when uncombined, persistently retains its solid condition, even in the hottest fire.

Bright light but no flame.

Remark, also, that this combustion is attended with a very bright white light, and compare it with the more violent combustion of hydrogen, with which most of my audience must be familiar.

Hydrogen burns with a flame, because it is a gas; but this flame is almost invisible, because gases, however intensely heated, do not emit a bright light. The charcoal burns without flame because it is a permanent solid, but for this very reason it emits a great amount of pure white light. So far, at least, as ordinary experience extends, white light is emitted only from ignited solid matter.* Therefore neither white light nor flame is a necessary concomitant even of the most rapid combustion, the first depending solely on the solid, and the last on the aeriform, condition of the burning substance. If, as in the burning of a candle, both flame and white light attend the process, it is because both solid and aeriform matter are there burning; and when we come to examine this phenomenon more closely, we shall find that the result is produced by a most beautiful adaptation of properties.

Let me next call your attention to the importance of the infusibility of charcoal in connection with its use as fuel. However high the temperature at which it burns, however intense the furnace heat, charcoal never

Use as fuel depends on its infusibility.

* I use the phrase *white light* because an ignited gas or vapor may emit a colored light, and it has been found that the color is in each case determined by the chemical composition of the ignited mass; but the light emitted from the gases or vapors of which the flames of ordinary combustibles consist, is at best very feeble. The light of such flames, as will soon appear, comes almost entirely from solid particles of charcoal, and when these from any cause are not present, the flame only yields a very faint blue light. The appearance of the whole flame is then the same as that which may always be seen near the orifice of a bat-wing gas-burner.

loses its solid condition, and on this wholly depends its application for generating heat. Were coal fusible, even at a very high temperature, it would melt and run out from our grates and furnaces, and the genial fire could not as now have been localized on the hearth. The enjoyment of the social fireside is thus closely connected with a familiar property of this wonderful element.

The products of combustion aeriform.

But our fire is slowly burning away, and already more than one half of the coal has been consumed. What has become of it? Do you point to the ashes? These are only the earthy impurities, which are more or less mixed with the pure carbon, and constitute but a small fraction of the whole mass of the coal. The carbon itself has combined with the oxygen of the air and formed a colorless and invisible gas, which has escaped by the chimney. I have already stated in my Lecture on Oxygen, that this gas has acid qualities, and we have named it carbonic acid. Reflect now on the importance of the circumstance, that this compound of oxygen and carbon is aeriform, and consider what a marked evidence of design and adaptation is to be found in the very fact that the products of ordinary combustion are invisible gases, which ascend our chimneys and are wafted away by the currents of the atmosphere. As the common experience is confined to the burning of coal, wood, oil, and similar combustibles, consisting mainly of carbon and hydrogen, men naturally associate with smoke the idea of a gas, and are apt to think that the aeriform condition is a necessary result of the nature of things. But it is not so. This peculiar provision in the case of carbon and hydrogen is an exception to the general rule. The two combustible elements which are most closely allied to carbon in all their properties, boron and silicon, not only form solids by burning, but two of the most fixed solids known in nature, one of which, silica, constitutes, as we have seen, at least one half of the rocky crust of our globe; and the same is true of almost all the other combustible elements. A very interesting experiment in illustration of this fact may be made by burning a piece of phosphorus under a dry glass receiver. The smoke of phosphorus is solid, and it will fall in thick white flakes, producing within the glass the appearance of a miniature snow-storm. Picture to yourself the desolation which would be produced were the order of nature so far changed as to make the products of burning coal like those of burning phosphorus. Every furnace would become a volcano, and we should soon be buried beneath the smoke of our own fires. When, now, we consider that a special provision has been made in the case of that substance whose combustion administers to our wants by evolving light and heat, what evidence does it open to us of the all-wise forethought of the Great Original!

But this is not all. Let me now call your attention to an additional fact in regard to the carbonic acid which is escaping from our coal fire. The gas is entirely devoid both of odor and of taste, and, moreover, when in a sufficiently diluted condition it can be breathed with impunity. Consider what an amount of this product is daily formed, and you will then be able to appreciate the importance of this circumstance. The amount of carbonic acid which escapes from an average-sized iron blast-furnace in the course of a single hour is equal to at least two tons, and the amount which is generated even by our coal fire is surprisingly large. Moreover, no less than two thousand tons * of this gas are breathed into the air by the population of this city in a single day. If carbonic acid had been a pungent or corrosive gas, coal could not have been used as fuel; for its combustion, like that of sulphur, would soon have rendered the air irrespirable. But so entirely destitute is it of any perceptible odor or taste, that, although it has been evolved in these immense quantities from every fire lighted by man since he appeared on the globe, it so entirely escaped notice that its existence was not even suspected until it was discovered by Dr. Black about a century ago.

The products of combustion harmless.

There is still another remarkable phenomenon attending a coal fire, which, although it cannot be made evident to the senses, has been substantiated again and again by the most accurate experiments. The volume of carbonic-acid gas formed by the combustion is exactly equal to the volume of oxygen con-

The smoke has the same volume as the oxygen consumed.

* Calculated for 400,000 inhabitants.

sumed. It is a consequence of this fact, that the volume of the air is not in the slightest degree increased by the vast quantity of carbonic-acid gas which is daily poured into it. The gas occupies precisely the same space as the oxygen removed during the combustion, and thus the equilibrium of the atmosphere is not disturbed. It is true that we probably cannot see all the bearings of this simple provision; but we know enough to recognize in it a most marked evidence of design.

The last fact in connection with the coal fire to which I would direct your attention is the large amount of heat which the combustion of coal liberates, and on which its use as fuel very largely depends. One pound of charcoal, in burning completely, generates sufficient heat to raise the temperature of 80.8 pounds of water from the freezing to the boiling point. Every pound of charcoal may, therefore, be regarded as containing sufficient heat to boil eighty pounds of ice-cold water. What a vast amount of heat then lies buried in those inexhaustible beds of mineral coal, in which our country is so rich! And have we not another remarkable evidence of Divine wisdom in the fact that carbon, a substance which, on account of its infusibility and other qualities, is so well adapted for fuel, has been made a great reservoir of heat, from which man can draw an unlimited supply? When we remember that this heat, through the expansion of steam, may be converted into mechanical force, and that hence these beds of

Heat of combustion.

coal are not only magazines of heat, but stores of force, which have been accumulating from the foundation of the globe for the use of civilized man, and when we reflect that it is this force which is animating our commerce, weaving our cloth, forging our iron, and impelling the printing-press, how can we express our praises of the foresight of that Providence which endowed coal with such wonderful qualities, made it a vast repository of heat and of force, and then spread it bountifully over the globe?

We have discovered, my friends, all these wonderful indications of design and adaptation in this simple experiment, so familiar as to be almost trite, so frequently repeated as to pass unnoticed, and they are constantly speaking to us of the great Author of nature from the fireside of every home, and from the furnace of every work-shop in the land. The followers of Zoroaster still worship, in India, fire as divinity, and regard these burning coals as sacred. Behind this superstition and idolatry there is concealed true wisdom, by which we may well profit. Fire is neither divinity, nor yet its emblem. It has no other reality than as a phenomenon attending a chemical change; but in the qualities with which charcoal has been endowed in order to produce this phenomenon, in the delicate adjustment of forces by which the destructive change is confined within due limits, there are indications of divinity which may well make us thoughtful, and consecrate with additional sanctity the family hearth; and if I have succeeded, however imperfectly, in making audible to your intellectual ear this mute eloquence of burning coal, our time has not been spent in vain.

I have thus far drawn all my illustrations from the burning of charcoal and hard coal, simply because these familiar forms of fuel are nearly pure carbon, and the phenomena attending their combustion are comparatively simple. They burn, as we have seen, without flame, and for the reason that carbon does not volatilize, even at the highest temperatures. It is different, however, with soft coal, wood, oil, wax, and all other combustible materials which are used for generating light. They do not consist wholly of carbon; but this element is always combined with hydrogen, and most of these combustibles contain also in addition a limited amount of oxygen. When heated, they all evolve common illuminating-gas, and for this reason burn with a flame. In fact, the gas we are burning here to-night was made from just such materials. If you will visit the gas-works of this city, you will see long rows of iron retorts, firmly built into large brick furnaces. In these retorts the gas is made, and they are connected by means of a complicated system of tubes with all the numberless gas-burners of this large city. Every few hours the retorts are charged with soft coal, which soon becomes heated to a low red heat. At this temperature it slowly gives off gas, and it is the gas thus

Burning of soft coal.

Manufacture of illuminating-gas.

formed which is now illuminating this hall. After three or four hours the gas has been all driven off, but there is still left in the retorts the greater part of the carbon of the coal, in a condition which is called coke. This is then removed and used for feeding the furnaces, and a new charge of soft coal is introduced in its place. Coke is an excellent fuel, but, like charcoal, it burns without flame.

The processes which, in the manufacture and use of illuminating-gas, are spread over a whole city, are united in every soft-coal fire. The gas which is burning at this jet was generated in the retorts of the gas-works, and brought here in iron tubes to be burnt. In the grate the gas is made and burnt in successive moments, but the process is identical in both cases. When you throw a fresh supply of soft coal on the grate, the first effect of the heat is to generate illuminating-gas, which at once takes fire and burns with a brilliant blaze. But after some time the flame ceases, because all the volatile elements of the coal have been expelled, and the coke, which is left, merely smoulders, like charcoal or anthracite. What is true of soft coal is also true of wood and of all this class of combustibles.

Flame, as I have before stated, is in all cases burning gas. As we are generally familiar with it, flame is a cloud of illuminating-gas combining on its exterior surface with the oxygen of the air. In a gas lamp the gas is supplied ready made at the jet. In an oil lamp or a

Nature of flame.

candle, the gas is manufactured as fast as it burns. The use which we make of the flame in all these cases, is to generate light, and the qualities of carbon have been most admirably adjusted to produce that result. This is the point which I wish next to illustrate, and we shall understand this beautiful example of adaptation more readily by analyzing the burning of some one of the light-generating materials. I will therefore select a common wax candle as my example, because it is familiar to every one, and illustrates all the points I have in view.

Nothing could be simpler than the candle itself. It is a long cylinder of wax formed around a string made of loose cotton threads, which we call the wick. The wax, that familiar secretion of the honey-bee, is composed chemically of carbon, hydrogen, and a little oxygen; the wick, as the microscope would show us, is merely a collection of fine vegetable tubes. Let us now light the candle. For that purpose we apply the flame of a friction match to the end of the wick, and mark the result. The heat of the match melts the wax around the base of the wick, and now the peculiar virtue of these vegetable tubes comes into play. All fine tubes have the power of sucking up liquid, and the finer the tube, the greater the height to which the liquid is thus elevated. The tubes of the wick are vastly finer than any that man can make, and the melted wax is at once drawn up to the flame of the match. There it is volatilized by the high temperature, and a cloud of red-hot combustible gas forms around the summit of the wick. Like the rain-drop, or any other fluid body in a free state, it assumes a spherical form, but being much

lighter than the air, this sphere of gas no sooner forms than it begins to ascend, and, being very combustible, is burnt up by the oxygen of the air with great rapidity, so that before it has risen an inch from the wick it is reduced to a point. Meanwhile, however, the first sphere is followed by others, which in rapid succession meet with the same fate, and at any moment we have a large number of these little spheres, one above the other, rapidly diminishing in size from the lowest to the highest, which has then become a mere point. Hence results the familiar conical form of the flame. But our match is long since burnt out, and what, you will ask, now volatilizes the wax? Solely the heat evolved by the burning gas. This heat converts the wax into vapor as fast as it creeps up the wick, and thus the flame being constantly supplied with combustible gas, the candle continues to burn until it is all consumed. The candle-flame is then merely a cone of volatilized wax, rapidly combining on its exterior surface with the oxygen of the air, and as rapidly replenished from below by the constant conversion of fresh wax into vapor. In this process light and heat are evolved; but these are generated solely on the exterior surface of the flame, where the burning takes place. Within it is perfectly dark, as can be easily shown by pressing down upon it a piece of window-glass, through which the interior may be seen. Let us now study this chemical process more carefully, as it is on a very delicate play of affinities that the whole illuminating power of the flame depends.

The combustible gas formed from wax is composed essentially of charcoal and hydrogen. The light and combustible hydrogen has so great a tendency to retain its aeriform condition, that, when combined with carbon, it renders even this, the most fixed of all the elements, aeriform; but the moment the bonds of chemical affinity are loosened, the carbon resumes its solid condition. Such a change takes place in the flame, and it is the particles of solid charcoal thus liberated which render it luminous. Of the two elements of the gas, hydrogen has the greatest affinity for oxygen, and therefore burns first, momentarily setting free the carbon, which is sprinkled in a fine powder through the burning gas. This is at once intensely heated, and each glowing particle becomes a centre of radiation, throwing out its luminous pulsations in every direction. The sparks last, however, but an instant; for the next moment the charcoal is itself consumed by the fierce oxygen, now aroused to full activity, and nothing but a transparent gas rises from the flame. But the same process continues; other particles succeed, which become ignited in their turn, and hence, although the sparks are evanescent, the light is continuous.

Thus it appears that all our artificial light, the light which we are enjoying this evening, depends upon this provision, by which the particles of charcoal linger for a moment in the flame before they are burnt. Let me again repeat, white light is emitted only by ignited solid matter. The flame of pure hydrogen gives very little light, because there are in it no solid particles, and were the affinity of oxygen for carbon slightly greater than at present, the flame of the candle would be as little luminous: then the carbon would burn simultaneously with the hydrogen, and there would be no pulverized charcoal in the flame to radiate light. On the other hand, were the affinity of oxygen for carbon a little less than at present, the carbon particles would not burn in the flame, but would escape from it in clouds of dense soot. Our Heavenly Father has so carefully adjusted the relative affinity of oxygen for the two elements of these light-giving gases, that the hydrogen should burn a small fractional part of a second before the carbon. During this brief interval of time, imperceptible to our unaided senses, the solid particles of charcoal are set free, become ignited, and give motion perhaps to a single wave of light; but the instant after, they too rush into combination with the great fire-element, and not a particle is left to dim the transparency of the air. The smallest variation in either force would destroy the adjustment by which this result is produced, and our lamps and candles would cease to give their light. How delicate the adjustment! How beneficent the result! How evident the design!

To me the marks of God's designing hand are more conspicuous in that familiar candle-flame than in the grand cycles of astronomy, or in the wonderful mechanism of the human body. I return to it again and again with renewed confidence, and always find fresh satisfaction and increasing faith. There are many who believe with Laplace that this glorious system of suns and planets, with all its complex movements and adjustments, might be evolved out of a nebulous chaos by the sole action of the *primary laws of motion;* and now, after the great French mathematician has furnished a world to begin with, a modern naturalist asks us to believe that this hand of mine, with all its wonderful combination of nerves, bones, and muscles, was developed out of the claw of an animalcule, or some such thing, by what he calls the *law of natural selection;* and although these and similar theories may be held consistently with a belief in a Divine Disposer of the spiritual affairs of men, yet it is too true that to many of their advocates the order of nature signifies nothing higher than self-existing matter, directed by inexorable necessity. But no cosmogonist has been able to go behind the chemical elements, and until human philosophy can show how these forms of matter, with all the marvellous adjustments among their properties, have been evolved out of the "star dust" of the original chaos, or out of nothing, and can adjust by natural causes the delicate play of forces in that most familiar of all phenomena, a candle-flame, it will not be able to overthrow the evidence of design afforded by this genial winter-evening light; and the fact that these would-be world-makers explain most satisfactorily what men know least about, is, to say the least, not in favor of their theories. Yes, my friends, it is these most familiar evidences of design which are the most impregnable against the attacks of materialism. It is these household altars that we find always burning to enlighten our dull understanding, to disperse our gloomy doubts, and to reveal to us the presence of our God.

Smelting of the metals.

The delicacy with which the affinity of oxygen for carbon has been adjusted appears still more wonderful when we consider another of the uses of this force in nature. The useful metals, which may be said to be the tools of civilized life, are seldom found in nature in a pure state. They generally occur combined with oxygen, and this compound, which is called the ore of the metal, is found in beds or veins of the rocks, where it has been deposited through the agency of water. After the miner has dug out the ore from the earth, and washed it free from impurities, it is the business of the smelter to melt out the pure metal. Now in this ore the metal is combined with oxygen, and unless the smelter could break this bond, the highest temperature of his furnace would be unavailing. But the merciful Parent of mankind, when he thus locked up these his choicest gifts, gave to man a key which would unlock the treasure-house, but left him to find out its use; and as in the progress of humanity the metals were required to advance civilization and multiply the comforts of life, the secret was discovered, and the treasures one by one were brought to light. The only key needed was charcoal. The Creator has endowed carbon with a power so strong, that it readily overcomes the force by which the metals are united to oxygen, and by simply heating the ore with charcoal the metal is set free. Would that I could give you an idea of the strength of the force which is required to produce this result. The affinity of carbon for oxygen is one of the most powerful forces known in nature, so great as to be immeasurable by our ordinary human standards, and yet it is this same force which produces that delicate result, the light of a candle-flame. With such wonderful skill does God wield these mighty agents of his power.

Consider, finally, how this power of reducing the metallic ores has been united in charcoal to those other qualities which render it so valuable as fuel. The smelter heats his furnace with the self-same coals which reduce the ore. These coals remain unchanged in contact with the ore until they have done their work, and then are converted into a colorless and harmless gas, which escapes by the chimney and is wafted away by the air; while, on the other hand, the melted metal, freed from its long imprisonment, flows out below in glowing streams, ready to be cast into thousands of useful forms.

Review, now, for a moment, the qualities of carbon, and notice how manifold and important are the functions which this element has been appointed to subserve. It has been made hard and brilliant, for the glazier's diamond and the monarch's crown. It has also been made soft and black, for the artist's pencil and the printer's ink. It has been made indestructible by atmospheric agents, and thus has preserved for us the wisdom of past ages, and will transmit our bequests of knowledge to those that are to come. It has been made combustible, and at the same time infusible, in order to localize our fires and confine them within their appointed bounds. It has been made a great reservoir of heat, in order that it might protect us from the winter's cold, and shed its enlivening warmth around the family hearth. It has been endowed with a strong affinity for oxygen, in order that it might reduce the metallic ores; but at the same time this affinity has been so carefully adjusted that the carbon particles linger in the flame for a moment before passing into invisible gas, and thus become a source of light as well as of heat. Lastly, the product of its combustion is a gas so transparent that it does not even cloud the atmosphere, and so bland that it bathes the most delicate organisms without harm. What an array of evidence have we here! But this, my friends, is only the first stage of that grand circulation of carbon in nature, which we proposed to ourselves as our subject this evening. The product of all these various processes of combustion is carbonic acid, and let us now follow this gas into the atmosphere, and examine some of its more familiar qualities.

Properties of carbonic acid.

Carbonic acid is so perfectly transparent and so devoid of every active quality that its presence cannot be recognized by any of our senses, and we must therefore call in the aid of experiment to make evident its existence. This is the reason why it remained so long unknown, the method we now use for detecting its presence having been first discovered by Dr. Black only a little more than a century ago. The method is very simple. Carbonic acid has a great tendency to combine with lime, and the result of this combination is that familiar white solid called chalk. Now lime is to a certain extent soluble in water, while chalk is insoluble; and hence, if lime-water is exposed to an atmosphere containing carbonic acid, the formation of particles of chalk, rendering the transparent solution turbid, will indicate the presence of the gas. Such a result is actually obtained by exposing lime-water in a saucer for a few days to the atmosphere, and any one can convince himself by this simple experiment of the existence of carbonic acid in the medium around us, as well as in the air which is exhaled from the lungs.

Indeed, the breath is so loaded with this product of combustion that lime-water is rendered milky by blowing into it for only a few minutes. The quantity of carbonic acid in the atmosphere, however, is relatively very small, not amounting to more than a few ten-thousandths of its whole weight. It enters to a far greater extent into the composition of many rocks. All limestones have the same composition as chalk, and contain nearly one half of their weight of carbonic acid, rendered solid by the force of chemical affinity. These rocks, indeed, are the great reservoirs of this aeriform compound, and when you consider how widely the limestones are distributed, underlying whole districts of country, reaching down to unknown depths, and piled up into vast mountain-chains, you can form some appreciation of the extent to which carbonic-acid gas was used in laying the foundations of the globe.

When pure, carbonic-acid gas will instantly extinguish flame, and is perfectly irrespirable, causing the epiglottis to close spasmodically and producing immediate death by *asphyxia.* When so far diluted as to admit of being received into the lungs, it acts like a narcotic poison, causing drowsiness and insensibility, and this even when a candle will burn in the gas. Hence the accidents which not unfrequently arise from heating a sleeping-chamber with a portable charcoal-furnace, or when from an imperfection in the flue the smoke of the hot-air furnace escapes into the house. Carbonic acid is not, however, poisonous in the strict sense of that term. On the contrary, it is always present in the blood in large quantities, and with it bathes all the tissues of the body. The carbonic acid results, as we have seen, from that slow combustion constantly going on in the blood, by which the animal heat is maintained, and it is an essential condition of life that this product should be secreted from the body as fast as it is formed. If the atmosphere contains more than a small percentage of the gas, the process of secretion is arrested, and fatal results necessarily ensue.

The density of carbonic acid is much greater than that of either of the other constituents of the atmosphere, the same volume weighing one half as much again as common air. Indeed, it is so heavy that it can be poured from one vessel to another like water, and the immense volumes of carbonic acid which are constantly flowing from our lungs and furnaces would cover the whole surface of the earth with their deadly vapor, were it not that the Creator has provided, by those simple laws of diffusion, which we studied in a former Lecture, that this noxious gas should be dispersed as fast as generated, and so mixed with the great mass of the atmosphere as to be rendered harmless by extreme dilution. The unfortunate accidents which sometimes occur to persons who descend incautiously into cellars or wells, where the carbonic acid is generated more rapidly than it can be dissipated, constantly remind us that the existence of animal life on the globe depends upon this beneficent provision. The large kilns in which lime is burnt into quicklime are constantly pouring out streams of carbonic-acid gas, and many a poor houseless wanderer, attracted by the heat of the kiln, has lain down to rest in the stream, and slept to wake no more. Were the force of diffusion much less than it is, we should all be constantly exposed to a similar fate; and when we lie down at night, it is only this guardian angel which prevents the deadly fumes of our own fires from descending on our beds.

Carbonic acid is soluble in water, a given volume of this liquid being capable of absorbing its own volume of the gas, irrespective of the temperature or pressure. We should therefore expect to find carbonic acid in solution in all water exposed to the air, and in fact a cubic foot of river, lake, or ocean water generally contains a very much greater amount of this gas than an equal volume of the atmosphere. Water, when holding carbonic acid in solution, has its solvent power very greatly increased. It then dissolves in large quantities all the varieties of limestone, and even granite rocks cannot wholly resist its action; but these solutions, when exposed to the air, gradually lose the carbonic acid, and with it their solvent power, incrusting with calcareous matter the moss, the twigs, or the walls of caverns on which the liquid may chance to rest. It is the solvent power of such water, acting slowly through ages of time, that has hollowed out that immense cavern in the limestone strata of Kentucky, and it is from the solution thus made that those stalactitic ornaments have been formed which add so much to its beauty and interest. It is also this same agency which in other places has deposited beds of calcareous tufa over great areas, and cemented together loose sands into firm rocks; and finally, it is from the lime dissolved in the water of the ocean, that the crustacea form their shells and the coral polypi build their reefs.*

The origin of carbonic acid is the same in water as in air. In the water we have not, of course, active combustion; but this, as we have seen, is an insignificant source of carbonic acid when compared with the never-ceasing functions of respiration and decay, and these are as active in the rivers, the lakes, and the oceans as in the atmosphere. Moreover, the purpose which the carbonic acid subserves is the same in both cases, and this deserves, finally, our attentive study.

I have already intimated that carbonic acid is one of the few articles of which the food of plants consists. Let us trace, for a moment, the history of the plant. The seed containing the germ is placed in the soil. The genial warmth of the sun calls it into activity, and it shoots forth its small leaflets into the air. For a short time the small stock of starch and similar nourishment stored in the seed by a wise Providence serves for its sup-

* The whole peninsula of Florida has been in great measure built up by these little animals with the lime rock which the waters of the Mississippi pour into the Gulf, and which they have dissolved from the lime deposits of our Western States.

port; but this is soon exhausted, and for the future the plant must depend for its food upon the soil and upon the air. The articles which compose its diet are exceedingly simple. They are water, carbonic acid, and ammonia, substances always present in the atmosphere and in every fertile soil. As soon as the young plant has expanded its green leaves it absorbs these substances, partly through its rootlets from the soil, and partly through its leaves from the air. The leaf, a tissue of minute organic cells, is the laboratory in which, from these few compounds, are elaborated the different organs of the plant. The sun's rays, acting on the green parts of the leaf, give to them the power of absorbing water, carbonic acid, and ammonia, and of constructing from the materials thus obtained the woody fibre, starch, sugar, and other compounds of which the plant consists. We have analyzed the woody fibre, and we know that it is composed of charcoal and water. Nineteen ounces of wood contain nine ounces of charcoal and ten ounces of water. Moreover, the amount of charcoal required to make nineteen ounces of wood is contained in thirty-three ounces of carbonic acid. If, then, we add together thirty-three ounces of carbonic acid and ten ounces of water, and subtract from this sum twenty-four ounces of oxygen, we shall have just the composition of wood. This is what the sun's light accomplishes in the leaves of the plant. It decomposes the carbonic acid, and unites its carbon to the elements of water to form the wood.

What I have stated to be true of wood is equally true of starch, gum, sugar, and most of the products of vegetable life. All these, with a few exceptions, which I shall notice in the next Lecture, are prepared by the plant from carbonic acid and water, under the influence of the sun's light. Why it is that starch is deposited in the cells of the potato, sugar in those of the sugar-cane, and gum and woody fibre, more or less, in all plants, we do not know. These are the mysteries of organic life, which no science has been able to solve. This much, however, is certain. The acorn, buried in the ground, grows into the noble oak. Of that wide-spreading tree, at least nine tenths consist of charcoal and water. The water was absorbed, as such, directly from the atmosphere; the charcoal was recovered from the carbonic acid decomposed by the sun's rays. Here is the wonderful fact. The gentle influences of the sunbeam have the power of reversing the process of combustion, of overcoming the intense affinity of the fire element, tearing it apart from the carbon, and restoring it to the air. How great this power is, I have already endeavored to illustrate. I have stated that the affinity of oxygen for carbon is one of the strongest affinities known to nature, immeasurable by any human standard. In order to decompose carbonic acid in our laboratories, we are obliged to resort to the most powerful chemical agents, and to conduct the process in vessels composed of the most resisting materials, under all the violent manifestations of light and heat, and we then succeed in liberating the carbon only by shutting up the oxygen in a still stronger prison; but under the quiet influences of the sunbeam, and in that most delicate of all structures, a vegetable cell, the chains which unite together the two elements fall off, and while the solid carbon is retained to build up the organic structure, the oxygen is allowed to return to its home in the atmosphere. There is not in the whole range of chemistry a process more wonderful than this. We return to it again and again, with ever increasing wonder and admiration, amazed at the apparent inefficiency of the means, and the stupendous magnitude of the result. When standing before a grand conflagration, witnessing the display of mighty energies there in action, and seeing the elements rushing into combination with a force which no human agency can withstand, does it seem as if any power could undo that work of destruction, and rebuild those beams and rafters which are disappearing in the flames? Yet in a few years they will be rebuilt. This mighty force will be overcome; not, however, as we might expect, amidst the convulsion of nature, or the dashing of the elements, but silently, in a delicate leaf waving in the sunshine. And this is not all. Those luminous waves which beat upon the green surface of the leaf are there arrested, and their moving power so completely absorbed, that the reflected rays will not even affect the exquisitely sensitive plate of the photographer. But the power of the light has not been lost, and when the wood is burnt and the carbon converted back into carbonic acid, this power reappears undiminished in the heat which radiates from the burning embers. The heat, therefore, which the wood contains, and which it gives forth on burning, comes from the sun. What a beautiful provision of Providence have we here! During the summer, when the sun is warming us with his genial rays, he is also laying up in the growing wood vast stores of heat, which will warm us at the winter evening fireside, when his rays have been withdrawn.

But you will tell me, it is not wood, it is coal, which is burning in the grate, and you will lead me, perhaps, to the mouth of some black coal-pit, and ask if those dismal regions below ever saw the sun. Certainly! and it is one of the most remarkable revelations of modern science, that the stone-like coal was once alive. Coal is the remains of an ancient vegetation, which flourished on the earth ages before man first walked in Eden. The process by which it has been formed and buried in the earth is well known. You can see it now forming in many tropical swamps. There you will find a vast mass of vegetable matter, the result of a rank vegetation, gradually decaying under water. The land is gradually sinking, and as this bed of peat sinks with it, it becomes covered with mud and sand, which numerous streams are constantly washing into the swamp. This goes on year after year, century after century, age after age, until the bed is buried hundreds of feet beneath the surface. In the mean time the vegetable tissues undergo a sort of internal combustion, similar to that which takes place in a charcoal mound. Wood consists, you will remember, of charcoal and the elements of water. The oxygen which it contains reacts on the carbon and hydrogen. Carbonic acid and water are formed, which escape, while the rest of the hydrogen and carbon unite together to form the coal. The reaction is a true process of combustion, and the heat thus evolved aids the chemical change, and gives to the coal its baked appearance. This change it requires long ages to complete. Millions and millions of times has the earth repeated its annual revolution around the sun, and the whole external appearance of the globe has changed since those mighty forests grew, which have been petrified in the coal. But though such long intervals have elapsed, their history has not been lost. It has been written on the rocks, the mighty monuments of past ages. The geologists have read it, and we know with as much certainty the form of the leaves and the structure of the stems of those ancient trees, as we do those of the oak or the chestnut. We know also that every atom of coal which now lies buried hundreds of feet beneath the surface was once a part of the atmosphere, and that the heat which it evolves by burning was received from the sun, when the carbonic acid was decomposed by the light in the leaves of the ancient trees. Consider for a moment of what immense value to man are those beds of coal. Without them modern civilization would have been impossible. Remember that since the dawn of creation the sun has been employed in accumulating these vast stores of force, and thus preparing the globe for civilized man. We may admire the genius of a Papin and a Watt, who have told us how to use this force, and who have thus covered the ocean with steamships, and the land with railroads; but let us not forget that infinitely greater wisdom which saw the end from the beginning, and before the mountains were brought forth or ever the continents were formed, laid up the beds of coal in the early strata, and preserved them through the long ages of geological time until the earth was fitted to become the abode of man.

I have now glanced at some of the distinctive features of the great circulation of carbon in nature, and have endeavored to show that the sun's rays are the prime moving power of the whole. I trust that you have been impressed with the grandeur of its cycles, the delicacy of its adjustments, and the mighty power of that mysterious influence by which it is sustained; but above all may it be my privilege to have succeeded in making clear to your intellectual vision those marks of wisdom and of power which have been so visibly stamped upon this Divine economy.

LECTURE VII.

TESTIMONY OF NITROGEN.

In order to complete my very imperfect sketch of the wonderful adaptations which the various qualities and functions of our atmosphere present, I wish in my Lecture this evening to examine with you the properties of nitrogen, that aeriform element which is its chief constituent, making up no less than four fifths of its entire mass. This gas, although so seemingly inert, discharges functions no less important than those of oxygen to the well-being of man.

Nitrogen not widely diffused. It is not, however, like oxygen, an element widely distributed in nature, and entering as a chief constituent into the composition of the globe. The atmosphere is the only great reservoir of nitrogen, and to this and to the bodies of organized beings its presence is almost exclusively confined. It seems to be the essential element of all the higher forms of corporeal vitality, and it is frequently called the *zoogen*, or life-generator. By some mysterious process it is constantly being withdrawn from the atmosphere, and entering into the composition of the numberless living forms which clothe the earth with verdure and crowd it with animal life; but these forms soon pass away, and by the inevitable process of decay the nitrogen is restored to the great reservoir from which it was originally withdrawn. Science has not as yet been able to follow all the steps of this remarkable process, but, nevertheless, enough is known to show that the properties of nitrogen have been most admirably adapted to the numerous important ends which it has been appointed to subserve.

Nitrogen is then peculiarly the element of the atmosphere. It not only constitutes the greater part of the aerial ocean, but it exists there in a perfectly free and uncombined condition, and — with the self-limiting exceptions just noticed — is found nowhere else. Occupying so important a place in the scheme of creation, we should naturally expect to find in nitrogen a substance full of the highest interest. Yet nothing could be less inviting than its external properties. A permanent gas, even at the lowest temperatures, without color or odor, it is entirely devoid of every active property. It will extinguish a candle immersed in it, and will not sustain animal life: but these are merely negative qualities; for animals cannot live in an atmosphere of nitrogen, solely because it does not contain oxygen, and it will not support combustion because it is not endowed with active affinities. And in all other outward aspects nitrogen is equally inert. It exerts no action whatever upon the most delicate chemical compounds, and with a few unimportant exceptions will not enter into direct combination with any of the chemical elements. Consider also the

Negative properties of nitrogen.

nitrogen as it exists in the atmosphere. Although in immediate contact with the most violent of the elements, and exposed to its action when in its fiercest state, under the varying influences of light, heat, and electricity, yet no combination between the two results, except to a very limited extent, and under peculiarly oblique conditions. Through an ordinary iron blast-furnace there pass in the course of a single day many tons of this mixture of nitrogen and oxygen called air. The oxygen, as we know, causes the most violent chemical action; but although the nitrogen is brought into contact with the same intensely heated coal and iron, no combination, at least of any importance, ensues.

Compounds of nitrogen formed only by indirect processes.

Shall we then conclude that nitrogen is entirely unendowed with chemical affections, — that it is capable of forming no chemical compounds, and of producing no powerful effects, — that it is, in fine, a *mere* dead weight in the atmosphere, placed there, for the want of something better, to fill up the void and to give the required density, as a ship is frequently loaded with ballast when there is a lack of freight? Such is the conclusion to which the appearances would naturally lead, and such is the conclusion at which the chemists arrived in the early stages of their inquiry. Yet no conclusion could be more at variance with actual facts; for so far is it from true that nitrogen is the uninteresting substance which these negative qualities would seem to indicate, that there are but few elements which form a larger number of compounds, or which are endowed with more varied powers when the necessary conditions of combination are fulfilled. Nitrogen can be made to unite with the other elements only by indirect and circuitous processes. It is one of its most distinctive qualities to avoid direct combination; but when the necessary conditions are present, it surprises us by the readiness with which it combines, and by the great variety and remarkable character of the resulting compounds. When we should least expect it, we find, not single compounds, but whole classes, springing into existence, which, while they often defy our investigations by their protean character, excite our admiration by the simplicity of their constitution and by the beauty of the plan according to which they have all been fashioned. The points, then, which especially characterize nitrogen, and in which the evidences of design in its constitution are to be traced, are, first, its unexampled inertness when in a free condition; secondly, the variety and remarkable nature of its compounds; thirdly, the peculiarly oblique processes by which all its compounds are formed; and, lastly, their very great instability.

Nitrogen the ballast of the atmosphere.

Nitrogen may be very appropriately termed the ballast of the atmosphere, and this is undoubtedly the most obvious of its functions. Air, you will remember, is not in any proper sense of the term a distinct substance. It is a mixture of several substances, or rather there coexist around the globe at least three different atmospheres, one of nitrogen, one of oxygen, one of aqueous vapor, and perhaps we should add, as a fourth, one of carbonic acid, each with its own peculiar characteristics, and so entirely distinct that it would retain all its essential properties were the rest removed. Again, when studying in our fifth Lecture the general features of the great aqueous circulation on the earth, we discovered that the whole plan turns on the fact that the atmosphere of aqueous vapor is mixed with a large mass of other aeriform matter, which moderates all atmospheric changes and mitigates the violence of their effects. It also appeared in the third Lecture that the atmosphere of oxygen has been subjected to a similar restraint, and that the aroused energies of this terrible destroyer are most carefully tempered by great dilution. As the atmosphere is constituted, the oxygen cannot reach the burning combustible without carrying with it the whole mass of the surrounding air; but if this mass of aeriform matter were not present, the devouring element would rush upon its prey with a fury which nothing could withstand, and iron * would burn as readily as straw. Moreover, in several other connections we have shown that it is an essential condition in the scheme of terrestrial nature that the air should have its actual density. See now how beautifully all the conditions are fulfilled in the atmos-

* An iron watch-spring burns with the greatest readiness in a jar of pure oxygen gas.

phere. The proportion of oxygen has been most carefully adjusted to the necessities of animal life, and made so small that the violence of the fire-element may be restrained within due limits. The amounts of aqueous vapor and of carbonic acid have in like manner each been accurately adjusted to the purposes which it was appointed to subserve, and then, in order to make up the required density, a large mass of a perfectly inert gas has been added. Thus in the very inertness of nitrogen we find the most obvious evidence of adaptation. Its negative qualities are precisely those required in a substance which is designed to act as so much dead material, adding to the density of the atmosphere without interfering with the functions of its active agents.

Evidence of special adaptation.

Consider, also, how very greatly this evidence of design is enhanced by the fact that nitrogen is found only in the atmosphere and in the bodies of organized beings, into which it has been temporarily withdrawn. It is not, like oxygen, carbonic acid, or water, a main constituent of the globe, and cannot therefore be regarded, as the fatalists would have us believe, as so much material left over after the solid globe had been condensed by the molecular forces from a chaotic nebula. Nitrogen is not only exactly adapted to the functions it subserves in the atmosphere, but moreover these are its only uses, and I cannot see how it is possible to resist the conclusion that it was especially designed for the place it fills. That you may appreciate the strength of this evidence, let me illustrate the subject by an example from common life, which will be more to our purpose than a philosophical analysis of the argument itself.

It does not follow that the square granite blocks which form the greater part of the front of yonder magnificent warehouse, however well adjusted they may be, were actually cut with reference to this building, although the strong presumption is that they were. Nor does it follow that those highly ornamented window-caps and that elaborate cornice were originally designed for this particular edifice, although the presumption that such was the case is still stronger than before. Nay, more, it is not even absolutely certain that those skilfully carved ornaments which adorn the front, and are built into the walls, were originally intended to be placed where they are, although to doubt this conclusion would be the extreme of incredulity. I admit, it is barely possible that they were originally made for another building, rejected, perhaps for some defect, and afterwards put up here. But I will show you where there is an evidence of design in the building-material of this warehouse which you will be forced to accept. It is not conspicuous, and might be overlooked. Just here at the corner of the building there is a very peculiarly shaped block of stone. You never saw one like it before. This extraordinary shape was required by the peculiar form of the building lot and the position of the walls on the adjoining estate. The sides of the lot are not perpendicular to the front, and the block has been cut to the precise angle of the bevel, and at the same time exactly fits the adjacent walls. The conclusion that this block was designed for that place is irresistible. No sane mind would doubt it for a moment. I do not say there is not one chance in many millions, estimated on the doctrine of probabilities, that a block of this exact size and shape might have been found among the refuse stock of the stone-cutter's yards; but I do say, that, *in the absence of absolute proof to the contrary*, the certainty that this granite block was wrought with reference to the place it fills, and that the exact correspondence of its dimension and angles was the result of measurement, is as great as it is possible to attain by any process of reasoning short of a mathematical demonstration; moreover, it is as great as can be obtained in physical science, or in any department of human knowledge one step removed from the facts of consciousness or of observation.

The evidence that nitrogen was designed for the place which it fills in the atmosphere is vastly stronger than this. The force of the argument in the illustration just cited evidently increases very rapidly, the more singular the shape of the granite block, and the more accurately its form has been adjusted to the place it fills. Now nitrogen is as unique among the chemical elements, as water is among the compounds. Its external properties are so entirely different from those even of the class of elements to which it belongs, that chemists can hardly believe that it is a simple substance, and for the last fifty years have been vainly attempting to decompose it; but it has resisted all their efforts, and the more intimately they have become acquainted with its properties, the more singular and exceptional it has appeared. At the same time, while presenting these remarkable anomalies, nitrogen has been fitted to the unique place which it fills in the scheme of creation, with a nicety and precision which it is as much beyond our powers of thought to conceive, as it is beyond my feeble language to describe. It is not only that one or two of the corners of this block of nature's edifice have been bevelled to an exact angle, but it has been adjusted at every point to the ten thousand conditions of that complex structure which I have been describing, but how imperfectly! during this course of Lectures, with a skill immeasurably beyond all human art, and with an intelligence which "looketh to the ends of the earth and seeth under the whole heaven." If this is so, — and you will find that my guarded expressions fall far short of the truth, — why not use in these matters of faith the same common sense which we apply with so much success in common life, and which in our daily intercourse it would be nothing short of madness to disregard? We do not hesitate to trust the skill and honesty of a fellow-man, whom we not only have never seen, but even as to whose character our sole evidence is the most indefinite testimony. Why, then, not accept the precious and comforting truths of religion, and repose equal faith in the providence of your Heavenly Father, on evidence which, you must admit, is ten thousand fold stronger, and when you have everything to gain, and nothing to lose? Is it said, there is still room for doubt? Of course there is. God be thanked! there is no relation in life in which there is not doubt. Were there no doubt, there would be no faith, no trust, no confidence, no love; the heart would be absorbed in the intellect, religion would become an axiom, and morality a formula of mathematics. Use but one half of the observation, one half of the intelligence, which are never at fault in the business of life, and these marks of the Creator's wisdom and providence which lie all around us will become as evident as the sun. Act on this evidence, and the door of grace will be opened, new light will stream into the soul, and all nature will be seen radiant with a Father's love.

Compounds of nitrogen.

All this striking evidence of design and adaptation we have discovered in the most obvious of the attributes of nitrogen, — in those merely negative qualities in virtue of which it increases the density of the atmosphere without interfering with the functions of the active constituents. It would not, however, be in accordance with that economy of resources which we find everywhere in nature, that the uses of nitrogen should be limited to this single object; and after what we have already seen to be true in the case of oxygen, we shall not be surprised to find this singular element suddenly changing its character and appearing in a new condition. The second point, as you will remember, which I am to illustrate in regard to nitrogen, is the variety and remarkable nature of its compounds, as well as the singularly oblique processes by which they are formed; and, having examined the marks of design it bears in its first manifestation, let us now study the no less impressive evidence presented by the second. It would be entirely out of place in a course of Lectures like the present, to describe in detail any of the countless nitrogenized compounds which are known to chemistry, and it would require a long course of Lectures merely to illustrate the characteristic features of the great classes into which they may be subdivided. I shall be able only to glance at a few general facts which illustrate the point now under discussion, and also the part which nitrogen plays in organic nature.

Although nitrogen presents such an indifferent exterior towards the oxygen of the atmosphere, it can, nevertheless, be made to combine with it by resorting to certain oblique processes, and there may be thus formed no less than five different compounds. Every one is familiar with that highly corrosive liquid called nitric acid, and this is one of the compounds in question. Under certain conditions it is formed by the union of the oxygen and nitrogen which are mixed together

Nitric acid.

in the air. The only essential difference between the bland atmospheric air and this highly active chemical agent consists in the fact that in air the elements are only mixed together, while in the acid they are chemically combined. Were nitrogen to be suddenly endowed with the active affinities which from its position among the chemical elements we might naturally expect it to possess, then a large part of the atmosphere would be showered down upon the earth in floods of nitric acid. It is only the unexampled inertness of nitrogen which prevents the formation of this deadly compound. But although so corrosive when pure, nitric acid when immensely diluted is one of the few materials which nourish and sustain vegetable life, and therefore provision has been made that it should be formed in the atmosphere, but only under very restricted conditions and to a very limited extent. When electrical sparks are passed through a confined quantity of air, in the presence of some alkaline substance, such as potash, soda, or lime, a very partial combination takes place between the two elements, and an infinitesimal quantity of nitric acid is formed. So also when organic matter decays in the presence of these same alkalies, a similar combination, although to a very slight extent, results. Nitric acid is endowed with such violent affinities that it does not remain in a free state. It at once enters into combination with the alkalies, forming a class of salts, of which saltpetre is the best known example, and from those salts the common nitric acid is extracted for the uses of the arts. Nitrogen, you will notice, acts here very much like a self-willed child. All the powers of nature cannot compel it to combine directly with oxygen; but if you offer to it these alkalies as an inducement, and make your approaches sufficiently indirect, you can coax it to combine, and nitric acid is then formed. We do not understand how the peculiar conditions just mentioned conspire to produce the result; but the whole phenomenon seems to be mysteriously connected with ozonized oxygen, and is undoubtedly another phase of that obscure subject, *allotropism*, to which we alluded in a previous Lecture. See now how beautifully this attribute of nitrogen has been adapted to the conditions of vegetable life, and made the means by which the plant is furnished with one of the articles of its food. Every discharge of lightning is accompanied by a partial combination of the elements of the atmosphere, and the nitric acid which is thus formed and washed down by the rain-water serves to fertilize the soil and bring the growing corn to maturity. So in like manner, when life is extinct, and the organized forms are resolved into their original elements, the very process of decay causes a similar combination, and thus sweetens the flowers which spring from the grave.

But not only does nitrogen combine with oxygen. It unites also with hydrogen, that element which is the very antithesis of oxygen, and forms a most remarkable compound called ammonia.

This substance is the very reverse of nitric acid in all its chemical relations, but like nitric acid it is a highly active and caustic agent. I need not dwell upon this fact; for the common smelling-bottle has made every one familiar with this pungent substance. Nitrogen manifests the same indifference towards hydrogen that it does towards oxygen, and the two elements can be made to unite only by indirect processes, which are not well understood. The most important of these is the process of decay. This destructive change in all the higher forms of organized beings is attended with the formation of ammonia, and the same nitrogenized compound is a uniform result of the normal functions of animal life. You will not, therefore, be surprised to learn that traces of ammonia, as of nitric acid, are found in the atmosphere and in all rain-water. Indeed, it is generally supposed that the two are in combination, forming a salt called nitrate of ammonia, but the amount present is at best very small.

Ammonia.

Ammonia is thought by many to be a more important article of vegetable diet than nitric acid; but our knowledge of agricultural chemistry is very imperfect, and chemists are not agreed on many of the most fundamental points.* Still, as I have

* Since these Lectures were written, it has been stated by several investigators that the chief nitrogen compound in the atmosphere and in rain-water is *nitrite of ammonia*, which differs from the *nitrate of ammonia* mentioned above only in containing a smaller proportion of oxygen. Whether the last is also normally present does not yet appear, and to what extent the one or the other may be concerned in the processes of

before stated, nitrogen is an essential element of all the higher forms of corporeal vitality, and compounds like those we have been considering are the appointed channels by which it is introduced into the organization of the plant. Had these compounds been allowed to form to any extent in the atmosphere, they would soon have rendered the globe uninhabitable. It was therefore essential that nitrogen should be endowed with that unexampled inertness which it manifests in its gaseous state. But had not at the same time a power of combination, under certain restricted conditions, been granted, this chemical element would not only have been an isolated phenomenon in nature, an exception to its general laws, but its usefulness would have been restricted to the least remarkable of its functions. Unlike the results of human skill, this creation of Divine wisdom has been adapted

vegetable growth, has not been determined. From one point of view, nitrite of ammonia may be regarded as composed of nitrogen gas and water, and some chemists believe that it is formed by the direct union of these two substances, and that this union is favored by the processes of evaporation, combustion, and decay, which are constantly going on in the atmosphere. This theory is certainly supported by many facts, and those who hold it generally believe that nitrite of ammonia is the chief, if not the sole, source from which the plants derive their supply of nitrogen, while others attach only a secondary importance to the recent experiments. If the theory is correct, the formation of nitrate of ammonia—the presence of which in surface-water, and in the soil, under certain conditions, is beyond doubt—would be the natural result of the subsequent union of nitrite of ammonia (formed as just described) with the oxygen of the air; but, as intimated above, the whole subject is still very obscure, and from any experiments yet made we should not be justified in drawing definite conclusions.

to the most varied and apparently incompatible ends; and while in the atmosphere it is a mere dead weight, it is also the most plastic of the elements, is capable of entering into the most complex relations, and thus serves as the peculiar substratum of all the higher forms of organized being.

The last point I am to illustrate in regard to nitrogen is perhaps the most characteristic of its features, and it is one on which its relations in the scheme of organized nature very greatly depend. All the compounds of nitrogen are very unstable, and the slightest force is generally sufficient to overpower the delicate affinities by which the elements are held together, when the nitrogen at once returns to its home in the atmosphere. Although this inert element may be coaxed into combination, it never forms strong compounds. Its affinities, although so varied, are at best very feeble and delicate. It is always a weak timber in a chemical structure, and when this timber breaks, as it certainly will, sooner or later, the whole falls. You will need no further illustration of this fact than to be told that gunpowder, percussion-powder, and gun-cotton are all nitrogenized compounds, and owe their well-known properties to the weak affinities of this element. Nitric acid is only a little more stable than these explosive agents, and ammonia, although one of the most permanent of nitrogenized compounds, is still very easily decomposed. Passing next to organized substances, we find this distinguishing

Instability of nitrogenized compounds.

character still more conspicuous. As we have already seen, it is always the nitrogenized compounds which start the decay in vegetable or animal structures; and thus the great characteristic feature of all organized matter, its proneness to change and decay, nay, even death itself, is clearly foreshadowed in the properties of nitrogen. When the Creator first endowed this element with its feeble affinities, He also passed the doom of all living creatures: "Dust thou art, and unto dust shalt thou return."

Here I must leave this division of my subject. It would be highly interesting to study the innumerable phases in which nitrogen manifests itself in the world of living matter; to trace how, under the guidance of that mysterious principle of life, the most complex organic compounds are educed from such simple materials as water, carbonic acid, ammonia, and nitric acid; to follow these nitrogenized compounds through their varied history, from the time they are first generated in the plant until they are incorporated into the brain, the muscles, and the bones of man; to notice at every stage the same instability which so strikingly characterizes all the compounds of this singular element, capable of existing only under the continued influence of the vital principle, and, when that ceases to act, gradually degenerating and falling back into the simple products from which they sprang; but all such details would be incompatible with the plan of these Lectures, and must therefore be reluctantly passed by.

If, however, I have been able to place before you in a clear light the main features of this remarkable element, — its isolated existence in the atmosphere, its unparalleled inertness in the aeriform condition, its power of combination under restricted conditions, the great variety and complexity of its compounds, and, finally, their singular proneness to decomposition and decay, — it is all that I could expect. We have seen that in each of these respects nitrogen has been adapted with exquisite skill to the unique part which it plays in the scheme of the world; and this element, although outwardly so unattractive and dull, has borne the richest testimony to the wisdom, the goodness, and the power of God.

Circulation of nitrogen.

Having now become acquainted with the characteristic features of nitrogen, let us next consider the part which this element plays in that grand circulation of matter in organic nature, which has been already in part described. I have before stated that the plant is a true apparatus of reduction, in whose leaves carbonic acid is decomposed by the solar light. The plant absorbs carbonic acid partly through its leaves from the air, and partly through its roots from the soil. The sun's rays, acting upon the green surface of the leaf, decompose in some mysterious way the carbonic acid, overcoming the intense affinities of its elements, fixing the carbon, and setting free the oxygen, to be restored to the air. From the carbon thus obtained, and from the water, ammonia, and nitric acid, which are the other articles of its food, together with a few inorganic salts, the plant constructs its tissues. If in their production carbonic acid and water alone take part, there result such substances as woody fibre, starch, gum, and sugar, and of these nine tenths of all vegetable structures consist. If the nitrogen compounds are likewise employed in the process, there are formed, besides, such nitrogenized products as albumen, caseine, and fibrine. These last names may not be so familiar to you as the first, but you are equally familiar with the substances, and will recognize them at once when told that the white of an egg is nearly pure albumen, that cheese consists almost entirely of caseine, and meat of fibrine. Although these substances are best known to us as animal products, they are likewise found in all those vegetables which are articles of food. Albumen and caseine can readily be extracted from either peas or potatoes, and gluten, the substance which gives tenacity to flour-paste, has the same composition as animal fibrine.

The plant a producer of organic materials.

The animal, unlike the plant, has not the power of forming the substance of its tissues from inorganic compounds, but it receives them ready formed from the vegetable kingdom. It transmutes the vegetable products into a thousand shapes in order to adapt them to its uses, but its peculiar province is to assimilate and consume, not to produce. The nitrogenized compounds just referred to are the portion of its

The animal an assimilator and consumer.

food which supplies the constant waste attending all the vital processes. The non-nitrogenized starch and sugar, although they form the greater part of our food, are never actually incorporated into the tissues of the body, and, as we have already seen, are merely the fuel by which its temperature is maintained. The animal may either receive its nitrogenized food directly from the plant, as is the case with all herbivora, or only indirectly, like the carnivora; but in either case the origin is the same, and by the process of digestion these, originally at least, vegetable products are assimilated and converted into bones, muscles, or nerves, as the necessities of the animal may require. We find that during this process these substances do not undergo any fundamental change, but merely become parts of more finely organized tissues. We discover in the blood albumen and caseine, having precisely the same composition as that which may be prepared from potatoes, and the substance of the muscle does not differ essentially from the gluten of flour-meal.

Do not, however, suppose that the part of the animal is less noble than that of the plant. It is really much higher. We must be careful to make a distinction, too frequently overlooked, between the organized structure and the material of which it consists. There is the same difference here as between a house and the bricks of which it is built. It was formerly supposed that organic matter was formed under peculiar influences, and subject to special laws. But it is now known

Distinction between matter and organism.

that animal and vegetable substances obey the same laws of affinity as mineral matter, and the recent progress of chemistry has given us great reason to believe that we may be able one day to prepare all the materials of which plants and animals build their cells. Here, however, chemistry stops and creation begins. The great Architect of nature alone can fashion this dead material into living forms.* The vegetable kingdom is a great laboratory, in which the sun's rays manufacture from the gases of the atmosphere, and from a few earthy salts of the soil, the different materials which the organic builders employ. There the bricks are made, and from these the animal builds his bones and muscles. He does not make the bricks, but he does what is far more glorious, he builds with them his delicate frame, and as the work of the builder is higher than that of the brick-maker, so in the scale of being is the animal higher than the plant, and the more noble in proportion as its structure is more intricate and elaborate.

While the plant is a true apparatus of reduction, the animal is a true apparatus of combustion, in which the substances it derived from the vegetable are burnt and restored to the atmosphere in the form of carbonic acid, water, and ammonia, ready to

* I do not forget the alleged facts of spontaneous generation; but in all cases where living organisms have been produced without visible germs, there has been a falling back of organized material from a higher to a lower condition; and, if we except certain very untrustworthy experiments of the late Mr. Weekes, in no case has even the lowest type of an organic cell been produced from unorganized matter, unless through the natural processes of growth.

be again absorbed by the plant and to repass through the phases of organic life. Our bodies are furnaces, — furnaces continually burning, — whose fuel is our own flesh, and the smoke of whose fires is the food of the plant.

Motion in the animal the result of burning.

Every time I strike a blow a portion of the muscle is consumed, actually burnt up in producing the force. In every muscular effort I make, in every word I utter, in every step I take, a portion of the muscles concerned is burnt, and motion can no more be produced in the animal body without a combustion of its tissues, than it can be generated in a steam-engine without burning fuel under its boiler. As in the steam-engine the burning fuel is the source of its power, so in the animal body the burning muscle is the immediate cause of all its motions. I will to strike a blow, but it is not I that produce the motion. It is the muscle, and in the exertion the muscle is consumed. The muscle, however, does not originate the motion, any more than the fuel originates the motion of the steam-engine. The fuel, we have seen, does not originate heat. It is merely a reservoir of heat, and in burning it merely gives up the heat it once received from the sun. So the muscle is merely a reservoir of force, and in burning it gives out the force it contains. The force it contains it also received from the sun, when its substance was formed by the sun's rays acting upon the leaves of the plants.

What a wonderful revelation is this! Muscular

power originates in the sun. We do not create the force; we do not originate it; we merely excite it. The force which originally came from the sun lies dormant in the muscles until our will calls it into activity. Our bodies are machines, perfect machines it is true, but yet machines. Like all other machines, they merely transmit power, they cannot create it. They very closely resemble a steam-engine. As we must constantly feed the engine with fuel, so we must supply our bodies with food in order to repair the muscle which is burnt, and we can no more be said to originate that force which manifests itself in our bodies, than the stoker, who shovels the fuel into the grate, can be said to originate the force of the steam-engine. We are not our bodies, although we live in them, and direct their motions. They move by forces which emanate from a source far higher than us, and we stand in the same relation to them in which an engineer does to his machine. Certainly Lavoisier, the great father of modern chemistry, had caught a glimpse of the results which it has been left for more modern science to establish, when he wrote: "Organization, sensation, voluntary motion, life, only exist on the surface of the earth, and in places exposed to the light. It might be said, indeed, that the fable of Prometheus was an expression of a philosophical truth, which had not escaped the penetration of the ancients. Without light, nature were without life and without soul; a beneficent God, in shedding light over

Muscular power originates in the sun.

creation, strewed the surface of the earth with organization, with sensation, and with thought."

Although, my friends, it thus appears that our bodies are mere channels of force, machines whose motive power emanates from the great centre of the solar system, let us, while we recognize this startling result of science, remember the no less certain fact of consciousness, — that we are not our bodies, though we live in them, — that this conscious personality is something entirely apart from, and infinitely superior to, these corporeal atoms in which it is temporarily enshrined, surviving as it does all their changes. Let us also keep clearly in view the still more glorious truth, that this machine, with all its infinite capabilities of good and evil, is put entirely at our command; that not one conscious motion can take place unless we will it; and that this will of ours can set in action a chain of causes which no space can bound and no time can limit. Let us then well consider how great is the power which has thus been delegated to us, let us duly weigh the awful responsibility it involves, and so act that, when the Master claims his own, we may not be ashamed to render up the account of our stewardship.

Relation of mind to matter.

Moreover, although it is true that these bodies themselves are constantly dissolving into air, that the material atoms which compose them will in a few short weeks all be gone, and that there is nothing but the shadow of our forms which we can call our own, we must also re-

The vital principle.

member that there is a mysterious principle within, constantly renewing and reparing our wasting frames,—a cunning architect superintending a thousand builders who are constantly reconstructing, with materials prepared by vegetation, the bones, the muscles, and the nerves, as fast as they are wasted and consumed; making, in a most mysterious way, beyond all human comprehension, here the fibre of a muscle, there the filament of a nerve, here building up a bone, there uniting a tendon, fashioning each with scrupulous nicety, and fitting each to its place with never-failing skill. But no sooner is the work of the architect done, than another great power comes in to destroy it. The oxygen gas which the blood absorbs in the lungs and carries to the different parts of the body burns up these carefully elaborated tissues, converting them into carbonic acid, water, and ammonia, which pass into the atmosphere, from which they originally came. Life is, in fact, a constant struggle between the builders and the destroying element of the air; and when its short term is ended, and the builders cease because they are wearied and few, then "the dust returns to the earth as it was, and "the spirit returns unto God who gave it."

But let us not sorrow as those who have no hope; "for we know that if our earthly house of this tabernacle was dissolved, we have a building of God, "an house not made with hands, eternal in the "heavens." And cannot He who has clothed us with our earthly house provide for us a better and more enduring mansion? and are not all these wonderful changes in our present bodies a foreshadowing of the final consummation, when our earnest desire "to be clothed upon" shall be satisfied, and "mortality shall be swallowed up of life"?

Such is a very imperfect sketch of that great cycle of changes, of which all organic nature is merely a passing phase. Let us review for a moment its main features. When the foundations of the globe were laid, there were collected in the atmosphere all the essential elements of organized beings. From this inexhaustible storehouse the plant absorbs water, carbonic acid, and ammonia, which were placed there for its use, and which have been made to serve as its nourishment and food. It is the special office of the plants to elaborate from these few mineral substances, and a small amount of earthy salts, all the materials of organized beings. The animal receives these crude materials already prepared, and builds with them its various tissues; but no sooner are the cell-walls finished, and the structure ready to discharge its vital functions, than it is consumed by almost the very act which gave it life. The carbonic acid, water, and ammonia are restored to the atmosphere, and the cycle is complete.

Review.

Of this Divine economy the sun's rays are the great moving cause, and it is their mysterious power which is constantly reappearing in all the varied phases of organic life. And not in these alone; for, as we have seen, this same gentle influence keeps in motion the aerial currents which blow our ships across the ocean. It raises the water which turns the wheels of our factories. It drives the locomotive over the iron road, and impels the steamer through the waves. It roars at the cannon's mouth, and charges the grander artillery of the skies. There is no motion on the globe which cannot be traced directly or indirectly to the sun, and were his rays to lose their mysterious power, all nature would become silent, motionless, and dead.*

But in thus tracing to the sun all these varied phenomena, let us not forget that we have not yet found the great First Cause. The problem is not yet solved; the profoundest truth has yet to be told. This mysterious force, which the sun pours in ceaseless floods upon the earth,—whence comes it? You have already answered the question. The answer is on your lips. I have but to re-echo it, and how can I better do this than in the words of that blind poet to whom misfortune had revealed the true meaning of light!

> "Hail, holy Light! offspring of Heaven first born;
> Or of the Eternal co-eternal beam
> May I express thee unblamed? since God is light,
> And never but in unapproached light
> Dwelt from eternity, dwelt then in thee,
> Bright effluence of bright essence increate."

* For a most eloquent statement of the same truth I would refer the reader to the closing chapter of a recent work by Professor Tyndall of London, on Heat considered as a Mode of Motion. It is but justice to myself to state that the last half of the present Lecture was first delivered before the Lowell Institute, in Boston, in the winter of 1853–54, essentially as it now stands.

LECTURE VIII.

ARGUMENT FROM SPECIAL ADAPTATIONS.

I HAVE endeavored thus far in this course of Lectures to present a few of the prominent illustrations of the attributes of God, which have been discovered in the adaptations of the atmosphere to the conditions of organic life on the earth. We have read together one brief chapter of that evidence of design with which the book of nature is filled, and I cannot but trust that we have gained from our study nobler conceptions and more enlarged views of the wisdom, power, and goodness of our Heavenly Father. Every one who accepts the Bible as a divine revelation will rejoice to find how beautifully and how entirely the facts of science confirm its great fundamental truths. But have not these evidences of nature a greater value even than this? Do they not prove, independently of all revelation, the existence of a wise and omnipotent First Cause, at least so far as there is any moral certainty in the world? I am persuaded that they do, and I believe that they furnish the only logical ground on which a system of revealed religion can be based. In my introductory Lecture I stated that I preferred to discuss the adaptations of nature as illustrations of the attributes of God, rather than

Adaptations a proof of God's being.

as absolute proofs of His being; but now that we have surveyed the ground, let us consider whether they are not really moral proofs, with all the certainty that any proof not strictly a mathematical demonstration can give.

Character of the argument.

The argument from adaptation is one which addresses itself to every human being. It is suited to every intellect, and comes home to every man's experience. It is based on a principle of the human mind, — whether the result of experience or of intuition we need not inquire, — which compels it to infer design when it sees adaptation. Who doubts that the flint arrow-heads and stone implements found in New England, rough and misshapen as they are, were made by men? To question the universal belief would be regarded as little short of insanity. Why then not apply the same common sense to the interpretation of nature? The unlettered do, and believe, in their simple faith, that the feathered songster and the delicate flower were made by their Heavenly Father's hand. It is only those of us whose minds have been unsettled by the subtilties of logic who doubt, and, if we could analyze our doubts, I think they would be found, in most cases, to arise from a vague fear that, since nature stands on a level so much above man's experience, the ordinary principles of reasoning may possibly not apply, and we may be misled by apparent analogies. But why this fear? There is no essential difference between the adaptations found in nature and the adaptations made by men. Both employ means to attain some important result, and in many cases they secure the end by precisely the same means. The telescope and microscope are but reproductions of the eye, and, although very imperfect reproductions at the best, they imitate in all their essential features this beautiful optical apparatus of nature.* It is true that the adaptations of nature are vastly superior to the results of human skill, and it is also true that their origin is beyond our personal experience. We have seen the process of making a watch and the process of making a telescope. We know how the principles of both were discovered, and the whole subject lies within the range of our experience; but no man ever made or ever can make an eye, and the whole process of its growth and development is utterly beyond the range even of man's conception. All this is true; but if you reflect a moment, you will find that this is just what is to be expected, seeing that God is the Creator and we are His creatures, and so far from weakening, this very characteristic greatly strengthens the evidence. Moreover, it must be remem-

* The power which the eye possesses of adaptation to near and distant objects, combining the uses of the microscope and the telescope, and the capacity of self-adjustment, preserving always a perfect achromatism and freedom from spherical aberration, have never been reached in nearly the same degree by art. Moreover, in the eye this perfection is attained with a focal length of only half an inch, which vastly increases the difficulty. It is also a fact worthy of notice, that the improvements in optical instruments have preceded rather than followed the discoveries of physiologists, thus serving to explain the functions of the eye; and inventions like that of achromatic lenses, to which men have been led by theoretical study, have been found to be anticipated in nature.

bered that, if the design is of an infinitely higher order, the evidence of the design is infinitely more ample. A rude, misshapen image is a convincing evidence of human intelligence; but all nature with its numberless adaptations — from the properties of the crude elements up to the wonderful structure of the human frame — is given us as evidence of the wisdom of God.

The argument cumulative.

The argument from the adaptations of nature is of the kind we call cumulative. Its force depends on the concurrence of many and varied examples. It is not based on one isolated case of adaptation, or even on a thousand; but there is a host of conditions, which no man can number, each adjusted to each, and all bound together in one harmonious whole. Consider only the examples we have discovered in the very narrow field to which we have limited our study. How numberless are the conditions on which the harmonious working of the varied functions of the atmosphere depends! In the first place, there is the expansive tendency of the air, sustained by the solar heat, and restrained by the force of gravity, by which alone it is held to the surface of the globe. Then there is the density, exactly adjusted to the human organism, and depending on the measures and weights of the solar system. Next there are all the delicate relations to light, heat, and electricity. Passing to the separate constituents of the atmosphere, there is oxygen, with its three distinct modifications, endowed with fiery affinities, and yet so carefully guarded as to be a beneficent servant of man, intrusted with most varied and seemingly incompatible functions, and discharging each with equal fidelity and precision; next, there is water, nourishing all nature with its dews and rains, tempering the polar climates with the latent warmth of its genial currents, and protecting with its great frost-blanket the delicate plants from the winter's cold, — exceptional in all its qualities, and each adapted to some beneficent end; then comes carbonic acid, concealing in its transparent folds the solid frame-work of all organized beings, and the source of those priceless beds of coal, with their inexhaustible stores of heat and force; and lastly, but not least in interest or importance, there is nitrogen, so remarkably inert, and yet endowed with such varied affinities, forming such numberless compounds, and imparting to all such singular instability. As we thus hastily review the ground we have surveyed together, you will recall the numerous adaptations which we discovered while studying the wonderful cycles of change in which all these substances conspire, wheel revolving within wheel, and yet all moving with such delicacy and beauty of adjustment that no jar is felt through all this complicated mechanism, and not the slightest derangement occurs in any of its ten thousand parts.

Now the argument for design unfolded in this brief chapter of the book of nature comes home to us with the cumulative weight of all this testimony,

Perhaps plausible objections might be urged against individual examples of adaptation which have been advanced; but any one who questions the general fact must be prepared to disprove all. Were there but a single instance of adaptation, or only two or three, the sceptic might urge with a show of reason that this was the result of accident, — arose from the "fortuitous concourse of atoms"; but the examples of adaptations which we have discovered merely in the atmosphere, all interlacing with each other, and all working together in the general scheme, are by themselves alone so great a number that, if we take no higher ground than the mathematical theory of probabilities, the chances against the supposition that this system, even as we know it, was the result of accident, are almost infinite, and can be expressed numerically only when the sands on the sea-shore are counted. If such, then, is the weight of the evidence which the atmosphere gives, what must be the force of the argument in which all nature gives its united testimony? Truly, the number of atoms in the universe is not sufficiently large to express the probabilities against this forlorn hope of atheism!

But, my friends, the sceptic should be heard, and, having presented our side, let us listen to what he has to say in reply. The whole argument from special adaptations may be summed up in a few words. Within the sphere of human experience, adaptation proves the existence of an intelligence adequate to the conception and

Summing up of the argument.

execution of the design. We find in nature adaptations similar to the results of human intelligence, only of an infinitely higher order, and hence by analogy we conclude that these must have issued from an infinitely wise and omnipotent Designer. The argument assumes the reality of the human intelligence as consciously a power and an originator within its own sphere, and reasons from this to a similar conscious intelligence in the Author of nature. The argument assumes, also, the truthfulness of the human faculties as a source of knowledge, without which it is of course useless to reason at all.

Burden of proof is with the sceptic.

Now the adaptations of nature are facts which every one must admit, the sceptic among the rest. Moreover, he must also admit that the conclusion which we have drawn from these premises is the all but universal conclusion of mankind. Plutarch, writing eighteen centuries ago, without the light of the Christian revelation, bears this remarkable testimony to the universality of the religious idea: "If you go through the world, you may find cities without walls, without letters, without rulers, without dwellings, without wealth, without money, without theatres and manly sports; but there was never yet seen nor shall be seen by man a single city without temples and gods, or without prayers, oaths, prophecies, and sacrifices, used to obtain blessings and benefits, or to avert curses and calamities. Nay, I am of opinion that a city might be sooner built without any ground beneath it, than a commonwealth could be constituted altogether destitute of belief in the gods, or, being constituted, could be preserved." * The discoveries of modern travellers have not more fully confirmed the general truth of Plutarch's statement, than the experiments of modern socialists have proved the soundness of his opinion. No savage tribe has yet been found on which a belief in a higher power has not at least glimmered, and no community which has attempted to ignore religion has lasted a generation. The sceptic, then, if he rejects our conclusion, is bound to prove that the natural inference of man is based in error. If he sets aside the general rule of faith, and refuses assent to the universal creed,—"Quod semper, quod ubique, quod ab omnibus creditum est," †—he must explain, whatever theory he may adopt, how it comes to pass that all mankind have been duped, and all nature has issued in a lie. The burden of proof is with him, and how does he meet it? Generally, in one of two ways.

In the first place, he attacks the validity of the conclusion on purely speculative grounds, saying

Universal scepticism.

that adaptation is no longer an evidence of design when applied to subjects beyond

* Εὕροις δ' ἂν ἐπιὼν πόλεις ἀτειχίστους, ἀγραμμάτους, ἀβασιλεύτους, ἀοίκους, ἀχρημάτους, νομίσματος μὴ δεομένας, ἀπείρους θεάτρων καὶ γυμνασίων· ἀνιέρου δὲ πόλεως καὶ ἀθέου, μὴ χρωμένης εὐχαῖς, μηδὲ ὅρκοις, μηδὲ μαντείαις, μηδὲ θυσίαις ἐπ' ἀγαθοῖς, μηδ' ἀποτροπαῖς κακῶν, οὐδείς ἐστιν οὐδ' ἔσται γεγονὼς θεατής· ἀλλὰ πόλις ἄν μοι δοκεῖ μᾶλλον ἐδάφους χωρὶς, ἢ πολιτεία τῆς περὶ θεῶν δόξης ὑφαιρεθείσης παντάπασι, σύστασιν λαβεῖν, ἢ λαβοῦσα τηρῆσαι. Plutarch, Πρὸς Κολώτην, xxxi.

† Vincentius Lerinensis, written 434 A. D.

the range of all human experience. He may support this position by questioning, with Hume, the competency of the human faculties as a source of knowledge, or, like Comte, he may deny all final causes, and maintain that there is no evidence of anything behind the external phenomena of nature; but whatever form the scepticism may assume, the conclusion is the same, and the argument for design is ruled out as invalid.

Design not capable of demonstration.

With regard to this position I have only a few words to say. Design in nature, I admit, cannot be *demonstrated;* for the truths of natural religion cannot be evolved from a mathematical formula. The argument is based on analogy, and although the analogies are so close and so broad that to my mind they amount to moral proofs, and the conclusion appears as certain as any theorem of geometry, still I admit that, in the technical language of logic, the evidence is probable, and not demonstrative. But as a student of physical science, it is not my business to defend the credibility of the human faculties, or to discuss the doctrine of causation. The task belongs to the metaphysicians, and, as I stated in my first Lectrue, I shall not encroach on their peculiar province. Nor do I think it important to dwell on the value of analogical reasoning. Modern writers have not been able to add much to Bishop Butler's masterly discussion of the subject, and every man, however sceptical he may be in his speculative opinions, must admit, with the author of "The Analogy," that "probability is the very guide of life." One consideration, however, may be of value in answering objections; namely, that, since the difficulties which are found in natural theology reappear with equal strength in all departments of knowledge, no objections can be reasonably urged against the methods of the former, which apply equally well to our most familiar processes of thought. It may be fairly presumed that such objections are more apparent than real, and that they indicate not the inconsequence of our logic, but only the necessary limitations of our faculties.

Now analogy is not only the guide of common life, but it is also the basis on which physical science chiefly rests; and if this method of reasoning be disallowed, all the results of science beyond those of mere observation and demonstration must fall with it. It is frequently said, that scientific truth can be demonstrated, but religious truth must be accepted on faith; and in part this is true; but the statement is one of those loose sayings whose partial truth only renders the latent error more dangerous. No word is more frequently misused than *demonstration*. Technically, this term only applies to that form of absolute proof with which we deal in geometry or pure logic; but, popularly, a principle is said to be demonstrated when all that can be claimed for it is a high degree of moral certainty. In this double use of the term the error of the above statement lies, for it is made in one sense, and — frequently at

Province of demonstration.

least — understood in the other. Truth wholly new is never reached by demonstration; for demonstration cannot yield what is not already implied in the premises with which it starts. The truths of geometry and mechanics may be demonstrated; but then they are virtually contained in the axioms and definitions on which these sciences rest. All scientific *generalization* is based on analogy; and moreover, a great mass of the scientific truth which lies within the range of direct observation we owe to the same principle; for even here analogy directed the student to what he subsequently observed.

Analogy the chief guide in physical science.

Indeed, the great inspiring and directing power in the mind of the successful investigators of nature is the force of analogy. It is this which constantly leads them to pronounce conclusions unsound, although apparently sustained by experiment, and to accept others which are seemingly at variance with facts. It is this which leads them through long and laborious investigations to establish principles which they believe to be true, and sustains them in their efforts through successive failures to ultimate success. As indefinite and uncertain as the analogies of nature frequently seem to be, as unsatisfactory as they may appear to the great mass of mankind, and as impossible as it is to make them intelligible except to those already versed in scientific inquiries, yet the history of science shows that, when based on an extended knowledge and a mature experience, they very seldom lead astray.

The method of scientific discovery is frequently misunderstood, and the philosophy of Bacon, however important in correcting old abuses, has done not a little towards creating the misapprehension. Many persons seem to think that the author of the Novum Organum gave to man a rule, by which, with the aid of a sort of mechanical logic called induction, the laws of nature may be discovered, very much as a last is turned out by a lathe. Yet nothing could be further from the truth. So far as the observation of phenomena is concerned, — which must always be the occupation of the great mass of scientific men, — the methods are as mechanical as those of other learned professions, requiring chiefly a quick eye, a delicate touch, a ready perception, and most of all a clear head capable of discriminating between the accidentals and the essentials, which are always singularly blended in natural phenomena. But the great generalizations, which form the framework of knowledge, are not reached by rule; nor as a general thing are they in their inception of slow growth. On the contrary, they usually come like intuitions to the mind, with the rapidity of the lightning's flash, and it is frequently possible to mark the day and the hour when the revelation was made. But such revelations of scientific truth are vouchsafed only to those highly favored minds which through long study and patient investigation have been brought into perfect sympathy with the harmonies of nature; and if we

Method of scientific discovery.

First the conception.

analyze the conditions of the mental process, we shall find that these great discoveries are really the result of analogical reasoning.

But although the conception is thus sudden, the verification of the truth is frequently long and laborious. Great discoveries are not achieved in an hour or a day. Nature has so concealed her truths, and surrounded them by so many adventitious circumstances, that they can be disclosed to the world only after long and careful study. First comes the conception, afterwards the toilsome investigation by which it is proved that the facts of nature accord with the generalization. The investigation may lead to a great modification of the original idea, or may show that it must be wholly abandoned, and meanwhile another may have taken its place, to go through the same scrutiny in its turn; but still the conception which proves to be the law of nature is the real discovery. This, as we have seen, is the result of analogy, and most clearly vindicates the relationship of the mind of man to the Intelligence whence issued the Universe.

Afterwards the verification.

Every great scientific generalization will illustrate more or less clearly the principles here stated. It is true that many minds frequently concur in developing one grand idea, and the evolution may occupy so long an interval of time that the new truth appears to be the growth of an age, rather than the gift of any one man. Yet it is possible in almost every case to trace the successive steps of the discovery. This is

Illustration of the method.

especially true of the law of gravitation. Whether the first idea was suggested to Newton by the fall of an apple, it is not important to inquire; but the popular anecdote illustrates the nature of the original thought, which was undoubtedly sudden and intuitive, although, as Newton has himself expressly stated, it was the result of analogical reasoning. The conception once formed, the work of verification was long and laborious, and the results were at first so unsatisfactory, that Newton at one time abandoned his theory altogether, as unsupported by observation. It was not, indeed, until a new arc of the meridian had been measured by Picard in France — several years after the first conception — that the facts were found to agree even approximately with the theory, and astronomers have been occupied ever since in verifying the grand thought. The same general facts reappear in the case of the wave-theory of light, first conceived by Huyghens and subsequently verified by the successive discoveries of Malus, Fresnel, and Young; and we may lay it down as an almost universal principle, that scientific truth is discovered through analogy and verified by comparison with the facts of nature.

If now you will turn to the great central truth of natural religion, you will find that it has as good credentials as the best established laws of science. We have first the conception,— not only the conception of a few highly gifted minds, but the universal conception of mankind. We find afterwards this con-

The fundamental truth of natural religion rests on the same basis as the laws of science.

ception verified, — not only in the history of the race, but also in the experience of each individual man, and moreover the conception is apparently intuitive in every mind. Even if the sceptic denies the reality of both special and general providence, he must admit that, as the almost universal rule, both history and experience have only served to confirm and strengthen the religious idea.

We return now to the remark above quoted, better able both to appreciate the truth it contains and to unmask the fallacy it conceals. A large part of the results of science may be demonstrated, but only such truths as are already contained in the premises on which the demonstration rests are capable of this absolute proof; and these are in all cases reached by the human intelligence working on its own definitions and processes of thought, and this, too, even when the theoretical truth is afterwards found realized in nature. The highest forms of scientific truth are not capable of demonstration, and rest only on probable evidence, although the probability in their favor is so great as to beget the highest degree of moral certainty. In like manner, a great part of the truths of religion must be accepted on faith; but then the evidence in favor of the great fundamental truth of natural religion is as strong as the evidence for the law of gravitation, and the certainty is as great. Moreover, faith is not peculiar to religion. All our knowledge not the result of personal observation and investigation is held on faith, that is, on

Province of faith.

trust in other men, and absolutely all knowledge is held on trust in the authority of our own powers. Much of the knowledge which we hold without question, it is utterly beyond the capacity of our own intellects to verify, and moreover, no one doubts the existence of truths which now lie beyond the scope of the most gifted genius, but which hereafter may be attained by man. The scientific truths which it is not essential for us to know are left in the dark on purpose to stimulate study, and thus to educate the human race. Religious truths, on the other hand, it is essential for us to know, and, since they in like manner transcend our present powers, they have been specially revealed. We are called upon to accept them on sufficient evidence, and this is all that is meant by faith. Faith, then, is as truly a ground of belief in science and in common life as it is in religion, and it occupies a more important place in religion only because religious truth is itself so important, and so greatly transcends, in its essence, our limited human faculties.

Our reply, then, to the first position of the sceptic is this. Your objections apply as well to all knowledge as they do to religious truth, and, if you are consistent with yourself, you must reject the evidences of science as well as the evidences of religion. As we are not prepared to go this length, we shall with equal consistency hold to both. It is but justice to state that Hume, the most philosophical of the sceptics, pushed his speculations to their necessary consequences, and denied the existence of matter and spirit alike. But although from its very boldness difficult to refute, this form of scepticism is by no means the most dangerous; for in the present age of the world a system of philosophy is not likely to gain many adherents which, in the first article of its creed, utterly shocks all human self-conceit, by declaring that man neither knows nor can know anything with certainty.

Absolute materialism.

In the second place, the sceptic attacks the argument for design by setting up a theory of his own to explain the origin of the universe. He tacitly admits that the burden of proof is with him, and that, if he rejects the popular belief, he is bound to show how this cosmos might have been issued without intelligence and without a God. This he attempts to do, and the result is nearly as many theories as there have been strong scientific intellects in the world united with unbelieving hearts. To refute each of these theories in detail would be a labor like that of Hercules in slaying the Lernæan Hydra; for until Almighty Power shall sear the foul sore from which the whole brood proceeds, their unholy heads will start up more rapidly than they can be cut down. The most daring theories of this kind are those of the German materialists of the present day. As much as they may differ among themselves in regard to details, the boldest of these speculators agree in maintaining that absolutely nothing exists, or ever has existed, except matter and motion; that matter in its essence is uncreated and eternal; that motion is self-sustained; that mind is only a mode of motion, and that all the phenomena both of matter and of mind are the working out of an inexorable necessity. Hence they conclude that religion is a fable, and immortality a dream.

Here is atheism. This is the natural fruit of materialism; and we are glad that it has ripened, that men may see how disgusting and revolting it is, and how corrupt the tree must be which can bear such fruit. We are glad that men should know what must be the result of all their vain speculation and the seeking after false gods. The theory is perfectly consistent with itself, and an absolute necessity if nature be divorced from its Creator; for all philosophy has proved that either the theory of the Christian, or this theory of the materialist, with all its enormity, must be true. There is no half-way halting-place between. This course of Lectures has been a continued protest against the materialist's interpretation of nature, and I have not another word to add; for if a man wishes to believe that his purest loves and his holiest affections are only motions of brain-particles, nothing that can be said would have the slightest weight. If he has not already the refutation in his own consciousness of being, human power cannot aid him; no philosophy can extricate him from the slough. "Ephraim is joined to idols: let him alone."

It is seldom, however, that materialism shows its revolting features among us. It is too cunning and too cautious. It always appears disguised, and is for this reason far more seductive. It presents the attraction of great learning and of great apparent profundity, entangling many in its meshes before they are aware of their danger. It does not deny the reality of the human intellect, but, on the contrary, takes pride in its authority and power. It even admits the evidence of design, but at the same time insidiously undermines all religious belief; not so much, however, by what it declares, as by what it leaves to be inferred; not so much by the doctrines it inculcates, as by the spirit it keeps alive and fosters. In this refined form, materialism is by far the most prevalent phase of the unbelief of our time, and it is difficult to meet chiefly on account of its very vagueness and simulation. It lives almost entirely in the ever-changing theories and speculations of science, which it utterly misinterprets and misapplies, forgetting that they are merely provisional expedients, which the next wave of advancing knowledge may wash away. *Development* is the pet word of its philosophy, and it constantly aims to show how the whole scheme of nature, with all its adaptations, might have been evolved through the concurrent action of various unintelligent causes alone. As it attacks the argument for design on scientific grounds, it becomes the duty of the student of nature to expose its errors. It is, however, a most protean antagonist, and no sooner is it defeated in

Materialism in its more refined forms.

Development theories.

one form than it reappears in another. Every new development theory in any department of science furnishes it with fresh food. For a long time the famous nebular hypothesis of Laplace supplied it with abundant nourishment, and quite recently it has taken a fresh start, and grown most vigorously, on Mr. Darwin's very ingenious book entitled "The Origin of Species." But these are only two examples of a large number of similar works, which, being less able and less original, have had their day and been forgotten.

The danger of these works lies not so much in what they actually contain, as in their general tendency; not so much in the theories of their authors, as in the wrong conclusions which will inevitably be drawn from them, and to which in many cases they logically lead. Darwin, for example, professes to show that all the living forms of plants and animals, man included, have been during the geological ages slowly developed from a few germs, or possibly from only one, by the action of a principle which he calls the "*law of natural selection*," and he sustains the hypothesis by a most formidable array of experiments and facts. Such a theory as this, ingenious if not true, professing to explain one of the greatest mysteries, and presented in a fascinating style, finds its converts everywhere, and this, too, on grounds entirely independent of its scientific merit. That very same noble aspiration which leads men to imperil even life itself in investigating the secrets of nature,

The danger of such theories.

makes them also ready to lend a willing ear to any theory which professes to explain the mystery of creation. Hence the reason why works like the Vestiges of Creation, and those just mentioned, captivate and injure so many. If they merely stimulated curiosity, and led to study, no one could object to their influence, however erroneous he might think their philosophy. But, unfortunately, most readers, of whom it is no disparagement to say that they are not in a condition to weigh the evidence, accept the theory without examination, and, if sceptically inclined, their whole belief in an overruling Providence is shaken to its base.

It is in vain to urge that these theories may be consistent with a pure faith; for as long as they are not so regarded by the popular mind, — which invariably appeals to them as proofs of materialism, — the evil which they cause is not remedied. It may be said, and said with some justice, that a writer cannot be blamed for the abuse of his theory; but it must be admitted that the abuse is a great evil, and an author, if he be a religious man, is bound to guard against it by every means in his power. We should be very slow to charge any man with infidelity, for we know how often the human mind, in its eccentricities and inconsistencies, has united a true faith to the most sceptical and subversive speculations. But we do say, that the least a Christian philosopher can do for his faith is to give such a tone and spirit to his work as to render misinterpretation impossible; and

Duty of scientific men.

if he neglects to do this, he has no right to complain if his own opinions are misjudged. It is to no purpose to reply, that religion is out of place in a work on science; for the class of books to which we refer are not addressed to scholars, but to the general reading public, and, in some cases at least, they are appeals from the settled convictions of thinking men to the unformed opinions of the popular mind. A scientific man who presents his theories in an attractive form to the public, before they have passed the ordeal of criticism by his peers, appeals to a court which has no right to judge, and whose opinion would be worth nothing if given. Real and permanent injury may be thus inflicted, and an author has no right to take advantage of the ignorance of his readers. If he does this to gain a transient notoriety, it may be a pardonable error; but I cannot find language sufficiently strong to express my condemnation of those popular writers who insidiously undermine the faith of the community by imposing crude speculations on their ignorance.

I shall not attempt to discuss the intrinsic value of the various theories of development. They have all been judged in a competent court, and the almost unanimous verdict of scientific men has been simply "not proven." This is true even of the nebular hypothesis of Laplace, which many astronomers are inclined to accept as a convenient theory, a sort of scaffolding for their science, but as nothing more durable; and still more is it true of the different theories which from time to time have been advanced to explain the origin of species. Man will not be likely to disown his divine parentage until some at least of the links which connect him with the brutes can be pointed out. If any theory of development be true, there must have been an indefinite number of links between the noblest chimpanzee and the most degraded Hottentot, and yet the progressionists, Darwin included, have failed to discover a single one.

Development theories do not impair the argument for design.

But leaving to those who are competent judges the task of exposing the unsoundness of these speculations, let us inquire what bearing they have on the evidence of design. I answer, absolutely none. Assuming that Mr. Darwin could establish his peculiar theory in all its generality, — and I have no doubt that it has an element of truth, — it would not impair the evidence of design in the slightest degree, and the same is true of any development theory whatsoever short of absolute materialism. Those persons who imagine that they overthrow natural religion, fall into a capital error. It requires manifestly the same infinite intelligence to create a universe by a process of development as by a single creative fiat. Your belief that the beautiful piece of mechanism standing on your mantel-shelf was made by an intelligent man, would not be impaired if you were told that the artist was employed several years in its construction. The evidence of design in the clock is in its beautifully adjusted mechanism. The evidence of design in nature is in the wonderful adaptation of its parts. We can easily go back in the geological records to the time when the present order of nature did not exist, and the fact that the innumerable forms of organic life, with the adaptations of currents, soil, and climate essential to their being, have been developed out of the conditions which existed on the globe during the coal epoch, is no less an evidence of design, than the fact that the clock was developed out of the crude iron and brass used in its construction.

"We lament," says a recent writer, "to see the question between a sudden and a gradual genesis of organic types discussed on both sides — not, indeed, by the principals in the dispute, but by secondary advocates — too much as if it were a question between God and no God. In not a few of the progressionists the weak illusion is unmistakable, that with time enough you may get everything out of next to nothing. Grant us, they seem to say, any tiniest granule of power, so close upon zero that it is not worth begrudging; allow it some trifling tendency to infinitesimal increment; and we will show you how this little stock became the cosmos without ever taking a step worth thinking of, much less constituting a case for design. The argument is a mere appeal to an incompetency in the human imagination, in virtue of which magnitudes evading conception are treated as out of existence, and an aggregate of inappreciable increments is simultaneously equated in its cause to *nothing*, in its effect to the *whole of things*. You manifestly want the same causality whether concentrated on a moment or distributed through incalculable ages, only, in drawing upon it, a logical theft is more easily committed piecemeal than wholesale. Surely it is a mean device for a philosopher thus to crib causation by hair-breadths, to put it out at compound interest through all time, and then disown the debt; and it is in vain, after all; for dilute the intensity and change the form as you will of the Power that has issued the universe, it remains, except to your subjective illusion, nothing less than infinite, and nothing lower than divine."

So far as the genesis of nature is concerned, I am not sure that a development theory is not the more probable. At all events, it is a question of facts to be decided by scientific investigation, and by that alone. Let us insist that all theories of cosmogony shall be judged on their own merits; but let us also insist that they shall be kept within their own sphere, and not allowed to have a voice in questions on which they have absolutely no bearing. Again I repeat, no development theory can impair the evidence of design; for that evidence is based on facts wholly independent of any theory of cosmogony, and to which all theories must conform. If they do not, they will inevitably fall. The difficulty to my mind in Mr. Darwin's particular theory is not in its development feature, but in the fact that he refers the development to what I can understand only as an unintelligent cause. He may be able to give a different interpretation to his theory, and I greatly regret that he does not; but unless he can show that it is consistent with the action of an intelligent First Cause, it will soon be forgotten, like those that have gone before it. This is the criterion by which all such theories are finally judged, after the excitement of the controversy by which they are heralded has passed, and after the common sense of mankind has settled down upon its sober second-thought. That they have an injurious bearing while they last, is frequently more the fault of the secondary advocates than of the principals in the dispute, and we must not expect to cure the evil by indiscriminate censure or by social excommunication. So long as man thinks, he will speculate; and I rejoice that neither political nor ecclesiastical tyranny can touch this prerogative of free thought. The true remedy consists in exposing the fallacy of the shallow philosophy which is so ready to bring forward these crude speculations as proofs of materialism, and also in diffusing among educated people more spiritual views of nature and its laws.

To this subject I shall return in the next Lecture. But so far as the argument for design is concerned, all these considerations are unnecessary. The evidence is so ample and broad, that we can afford to waive all that part of it which has been called in question by the progressionists, without weakening in the slightest degree the force of the argument.

Before the first organic cell could exist, and before Mr. Darwin's principle of natural selection could begin that work of unnumbered ages which was to end in developing a perfect man, nay, even before the solid globe itself could be condensed from Laplace's nebula, the chemical elements must have been created, and endowed with those properties by which alone the existence of that cell is rendered possible.

The evidence of design in matter prior to any theories of development.

But although, for the sake of argument, we might yield to the progressionists all those examples of adaptation which they claim to explain by their theories, such a concession is really of no value. The parts of nature, as we have seen, are so intimately linked together that, if there is design anywhere, there is design everywhere; and as the structure of the human body is prefigured by the earliest vertebrate forms buried in the geological strata, so, and as unquestionably, the whole scheme of organic life is prefigured in the gases composing the atmosphere. If, therefore, I have proved that there is evidence of design in the constitution of the atmosphere, I have also proved that the whole scheme of nature is the result of Divine Intelligence, and that the great argument of natural theology rests on a basis which no present theories* of develop-

* We of course refer only to such theories and speculations as are based on observed facts; for no others are worthy of serious consideration. Science has not as yet gone one step behind the chemical elements, and until it has, no speculations in regard to a primordial condition of matter can have any bearing on our subject.

ment can touch. To show that there is evidence of design in these stones of nature's edifice has been the chief object of this course of Lectures. It has been my constant aim to set forth in a clear light the startling fact that the foot-prints of the Creator are nowhere more plainly visible than on that very matter which the materialists so vainly worship, and if I have thus been able to remove doubts from the mind of any honest seeker after God, I shall feel that my labor has not been lost.

But however earnest our purpose or sincere our convictions, the spectres of our doubts will sometimes return, and hover around these evidences of our faith. Treat them not lightly either in yourself or in those you love. Respect all honest doubts; for it is the noblest natures which feel them and suffer most. His must be a dull heart which is not sometimes appalled by the mystery of our being. Remember, however, that these doubts are from within, not from without. They are the offspring of your fears, and not of your science. The evidence is ample. It is more faith that you need. Fight then these spectres of your mind as you would the enemies of your peace, not with doubtful disputations, but with earnest thought and prayer, and Power shall be with you

> "in the night,
> Which makes the darkness and the light,
> And dwells not in the light alone."

LECTURE IX.

ARGUMENT FROM GENERAL PLAN.

It has been my object in the previous Lectures of this course to develop before you the great argument of Natural Theology as it is presented by the atmosphere. I have endeavored to show that there is abundant evidence of design, even in the properties of the chemical elements, and hence that the argument rests upon a basis which no present theories of development can shake. Having dwelt upon the argument from special adaptations at as great length as my time will permit, I wish in my Lecture of this evening to ask your attention to another class of evidences of the Divine attributes, which, although less conspicuous, may be even more impressive to some minds than those we have studied. The indications of an Infinite Intelligence are not only to be found in the adaptations of nature, but they also appear, and still more visibly, in the grand laws by which the whole material universe is directed.

I am well aware that the laws of nature, so far from being regarded as evidences of the existence of a beneficent God, are felt by many minds to be actual hinderances to their faith. They are thought to give to the whole scheme of nature a mechanical aspect, and to be inconsistent with belief in a superintending Providence. I also know that there are many scientific men who regard the laws of nature as the manifestation of blind physical forces, and who recognize a Providence, if at all, only in the very few recorded instances where the normal action of these forces has been averted by a special miraculous interposition. But even admitting this philosophy, still I think it will appear that these laws bear so conspicuously the marks of Intelligence, and are so analogous to the results of human thought, that we cannot resist the conclusion that they were originally, at least, ordained by an intelligent Creator, or in other words that the laws of nature are the thoughts of God. For myself, I regard the laws of nature as the most direct evidence possible of Infinite wisdom, and it will be my object to show that this opinion is sustained by the strongest analogies.

Definition of physical laws.

Regarded from a scientific point of view, physical laws are merely our human expressions of that order which we discover in the material universe. In its highest form, the law is capable of a precise quantitative statement, and gives the bases for mathematical calculation and prediction. Thus the law of gravitation enables the astronomer to calculate what will be the position of the bodies of the solar system at any future epoch, and to predict almost to the very second the exact time when an eclipse will begin, and what will be the precise path of its shadow over the earth. The greater part of the laws of nature do not, however, admit of precise mathematical statement, and are merely the expressions of the order which has been observed in the phenomena of nature, whether in respect to form, in respect to number, or in any other particular. It is convenient to distinguish these merely *phenomenal laws* from the higher class, which are usually called *dynamical laws;* but the distinction is an artificial one; for it is probable, at least, that in all cases the phenomenal laws are merely the phases of some higher dynamical law not yet discovered. Moreover, if we believe that all phenomena are direct manifestations of the Divine Will, then there is no law apart from God. His action is not necessitated or prescribed by any conditions, even although imposed by Himself. He is constantly acting in nature, consciously and freely; but He acts uniformly, consistently, and with a plan, because He is omniscient and omnipotent. Man acts with inconstancy, because he is a finite being, and must be guided by probabilities; but with God, who seeth the end from the beginning, there is no "variableness, neither shadow of turning."

The whole material universe may then be regarded as the manifestation of one grand comprehensive creative thought, which God is slowly working out in nature. To study this thought in all its details is the prerogative of man, and this study has been the appointed means of cultivating his intellect and elevating his condition. From time to time the more gifted students have caught glimpses of parts of the grand thought, and these glimpses we call laws; but even the law of gravitation, the most perfect of all, is felt to be but a partial truth, and all eternity will be required to comprehend the scheme. Let us now, in order to elucidate and confirm this simple doctrine, compare some of the laws of nature with the results of human thought, and, whatever may be our theory of causation, we cannot but be impressed with the striking analogy between the two.

The idea of symmetry.

The idea of symmetry is inherent in every human mind. It may be more or less cultivated by experience, but the germs of the idea are found even in the savage. However rude his condition, man is pleased with a symmetrical disposition of objects, and his taste is offended when the laws of symmetry are grossly violated, although he may have no name for the idea. Corresponding with this idea in our minds, we find symmetry everywhere in nature. The parts of an animal are symmetrically arranged round the body, and the leaves of a plant are symmetrically disposed around the stem, but nowhere in nature is the idea of symmetry so fully developed as in the mineral kingdom.

Crystals.

Almost every solid substance, when slowly deposited from a liquid or aeriform condition, assumes a definite symmetrical shape which is peculiar to the substance. Such symmetrical forms are called crystals, and the process by which they are obtained is called crystallization. Freedom of motion — such as the particles of matter have in the fluid condition — is an essential condition of crystallization. Moreover, as the substance becomes solid, the particles must have sufficient time to arrange themselves in accordance with the tendency of the molecular forces, and the longer the time occupied in the process of crystallization, the more perfect we find the crystals. The perfect crystals represent the natural condition of a substance, and the peculiar form is the most essential and characteristic of all its properties.

Crystals are always polyhedrons, that is, solids bounded by plane faces, and these faces are symmetrically disposed about certain straight lines called axes. Moreover, it has been observed that, although we might conceive of an infinite number of systems of straight lines, there are only six, or at most seven, such systems in nature, and all crystals, with possibly one exception, are formed by the symmetrical disposition of planes around one or the other of the six systems represented in Fig. 2. The first five of these systems have each three axes, and the last alone has four. In No. 1 the three axes are of equal length and at right angles to each other. This arrangement is called the *monometric system*, from two Greek words signifying *one measure*. In No. 2 the axes are all at right angles, as before; but the vertical axis is either longer or shorter than the other two. This is called the *dimetric system*, — two measures. In No. 3 the axes are still at right angles, but of unequal length. This is called the *trimetric system*, — three

Systems of crystals.

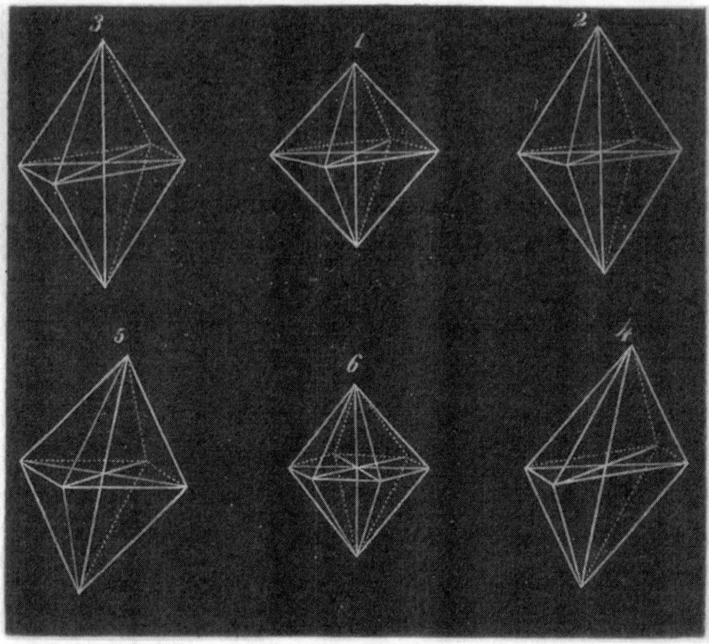

Fig. 2.

measures. In No. 4 all the axes are of unequal length, as in the last, and the two horizontal axes are at right angles. The vertical axis is also at right angles to one of the horizontal axes, but it is inclined to the other. Hence the system is called the *monoclinic system*,— one inclination. In No. 5 the three axes are all unequal in length, and all at oblique angles with one another. Hence the system is called the *triclinic system*,— three inclinations. Lastly, in No. 6, there are four axes; of these, three are horizontal, and form the diagonals of a regular hexagon, as shown by Fig. 3, and one is vertical, and at right angles

Fig. 3.

to the plane of the first three. This is called the *hexagonal system.* In all these figures the ends of the axes have been connected by lines, in order to render the character of the systems more evident.

It is now easy to see how far the idea of symmetry in man corresponds with the idea in nature; for, assuming these systems of axes as they are found, we may arrange planes around them in accordance with our ideas of symmetry, and then compare the results with nature. Let us now take the simplest of the systems, and follow out this idea. It will be evident on a moment's reflection, that the number of axes or lines of symmetry in a crystal must be at least three in number, and it will also be evident that the simplest arrangement of three straight lines is that represented in No. 1 of Fig. 2. This is called the monometric system of axes, and the lines, as stated above, are all equal, and all at right angles to each other. Moreover, by passing a plane through every two of these axes, as in Fig. 4, we may divide the space around them into eight similar parts, called octants. Let us proceed now to dispose planes symmetrically around these three axes, according to our own ideas of symmetry and entirely independent of any observation of nature.

Forms of the first system.

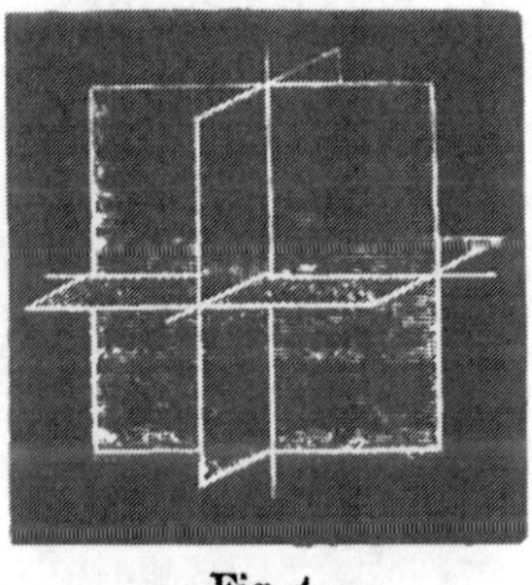

Fig. 4.

We may, evidently, begin by placing a plane at

the end of one of the axes, and perpendicular to it, as is represented in Fig. 5. But perfect symmetry will now require us to place a plane in a similar position at the other end of the same axis, and also at the two ends of each of the other axes. The result of such a disposition will be a solid bounded by six planes, called a cube. No. 2 in Fig. 12.

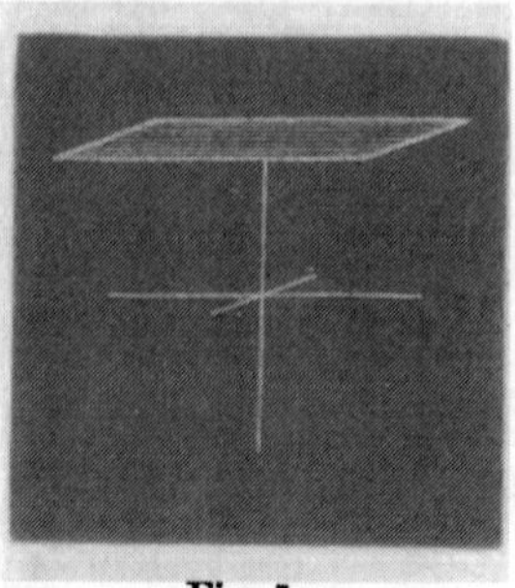

Fig. 5.

Next we may place a plane so as to touch the ends of two axes and to be parallel to the third, as in Fig. 6. But symmetry will now require us to place a plane in every other similar position, and the result of such a disposition will be a solid bounded by twelve planes, called a dodecahedron. No. 3 in Fig. 12.

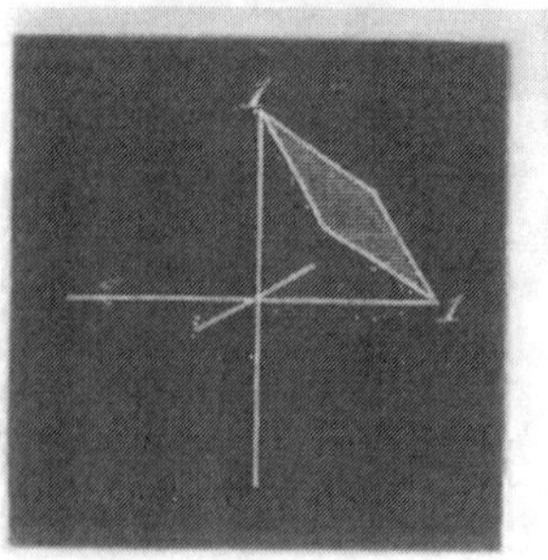

Fig. 6.

Again, we may place a plane so as to touch the ends of all three axes, as in Fig. 7. Symmetry will now require us to place a plane in every similar position, and the result will be a solid bounded by eight planes, called an octahedron. No. 1 in Fig. 12.

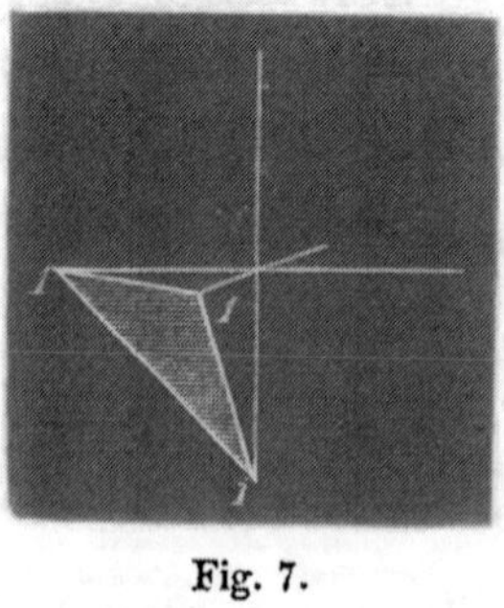

Fig. 7.

Then we may place a plane touching the end of one axis,

but so inclined as to intersect the prolongation of the other two axes at still greater, but equal,

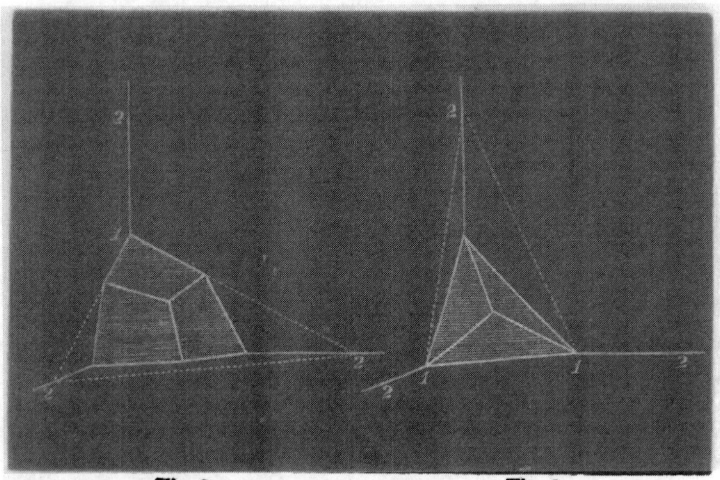

Fig. 8. Fig. 9.

distances from the centre, like the plane 1 : 2 : 2 in Fig. 8.* Symmetry will now require that we should place three such planes in each octant, making twenty-four planes in all, and the intersections of these planes will give the form represented by No. 4 in Fig. 12.

Further, we may place a plane touching the ends of two axes, and intersecting the prolongation of the third at any given distance from the centre, as 1 : 1 : 2 in Fig. 9.* Symmetry will now require that we should place three similar planes in each octant, twenty-four in all, and the intersection of

* These figures represent only one octant, see Fig. 4. The portion of the plane 1 : 2 : 2 in Fig. 8, and of 1 : 1 : 2 in Fig. 9, which appears on the resulting form, has been shaded, as have also the portions of the two planes which start from the corresponding points on the other axes of the same octant. The end of the axis is assumed to be in each case at the points marked 1.

these planes will give the form represented by No. 5 in Fig. 12.

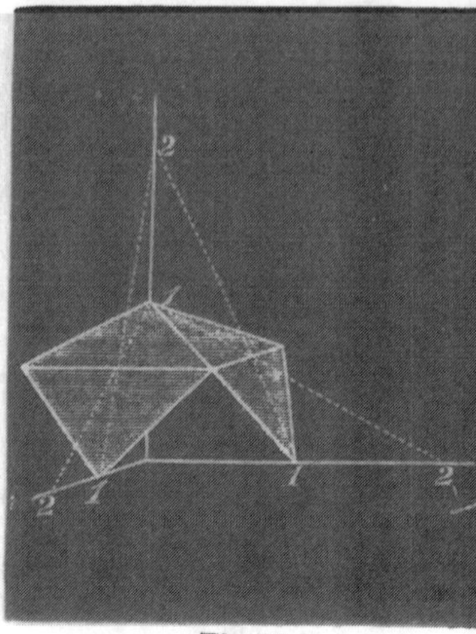

Fig. 10.

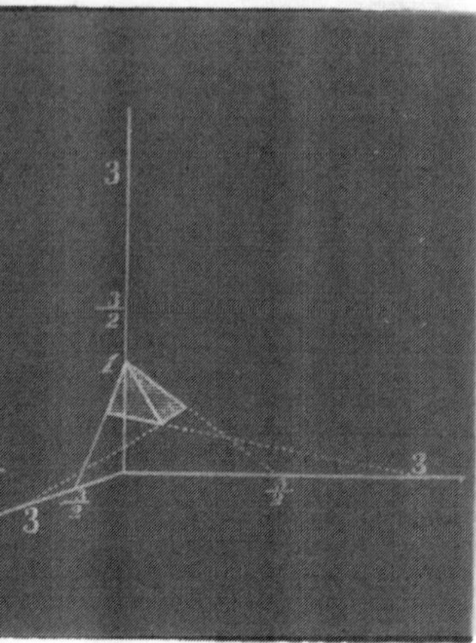

Fig. 11.

Also, we may place a plane touching the end of one axis, parallel to the second axis, and intersecting the prolongation of the third at some given distance from the centre, as either of the planes 1 : 2 in Fig. 10. Symmetry will require us to place a plane in every similar position, and we shall thus obtain the figure bounded by twenty-four faces, represented by No. 6 in Fig. 12.

Lastly, we may place a plane touching the end of one axis, and intersecting the prolongations of the other two at unequal distances from the centre, like either of the two planes $1 : 3 : \frac{3}{2}$ in Fig. 11. Symmetry will require us to place six such planes in every octant, — two at the end of each axis, — making forty-eight in all, and we shall thus obtain the form represented by No. 7 in Fig. 12.

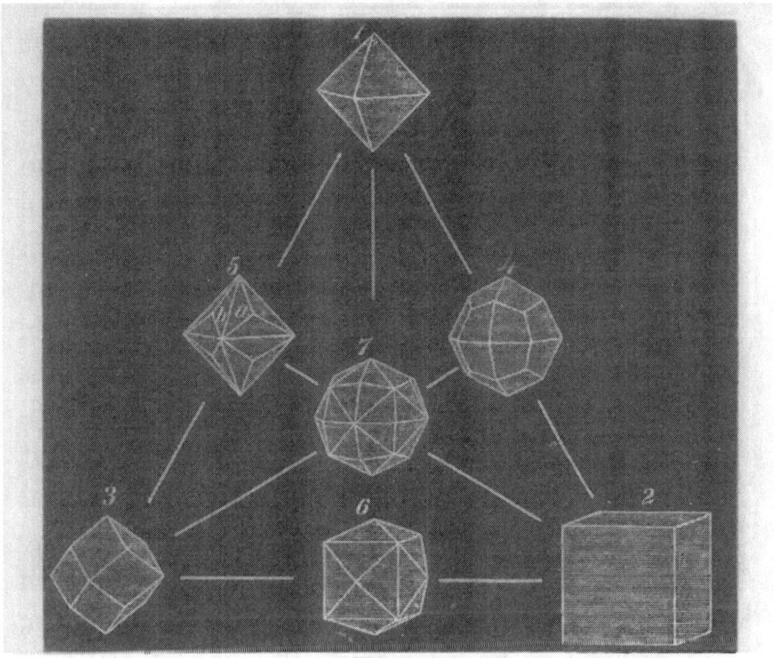

Fig. 12.

Of these seven forms, the first three, which make the vertices of the triangular group, are absolutely invariable. We may have cubes of different sizes, but the relative inclination of the planes is absolutely the same in all, and this is also true of the octahedron, No. 1, and of the dodecahedron, No. 3. It is different, however, with the other four forms. Here also the general disposition of the planes, and the general appearance of the form, are always the same; but the relative inclination of the planes may vary to a limited extent.

Consider, for example, the form No. 5 in Fig. 12. This form is composed of planes touching the ends of two axes and intersecting the prolongation of the third at any given distance from the centre, as shown in Fig. 9. Now it is evident that we may have an infinite number of planes fulfilling these conditions. The plane must touch the ends of two axes, and so far it is fixed; but it may intersect the prolongation of the third at any given distance from the centre, — at a distance of twice the length of the semi-axis, or of two and a half times the length, or of three times the length, or at any other distance. In this way its position may be varied infinitely; but whatever position is given to the first plane, all the rest must be perfectly symmetrical with it.

Consider, now, what change such a variation in the position of the planes will cause in the form. You will notice, by examining No. 5 in Fig. 12, that the form may be regarded as an octahedron, No. 1, bearing a triangular pyramid on each face. When every plane touches the ends of all three axes, the form becomes an octahedron. If the planes, still touching the ends of two axes, intersect the prolongation of the third at twice the length of the semi-axis from the centre, the form is that represented in No. 5. If they intersect at three times this length, we shall have a similar form, only the pyramid will be more elevated. If they intersect at four times this length, the pyramid will be still higher; and so on, until the planes intersect the third axis only at an infinite distance, in other words are parallel to it. Then the faces *a* and *b* on No. 5 will fall into the same plane, and we shall have the dodecahedron No. 3. Thus you see that Fig. 5 is an intermediate form between the octahedron No. 1 and the dodecahedron No. 3, shading into them on either side.

In like manner No. 4 is a variable form, intermediate between the octahedron and the cube. So also No. 6 is a variable form, intermediate between the dodecahedron and the cube; and finally, No. 7, the central form, is intermediate between all the rest, as indicated by the lines of the figure.

Such are the seven forms which we have been able to obtain by arranging planes around the regular system of axes, according to our idea of perfect symmetry. Moreover, since these seven forms, with the variations above described, are the only forms which we can obtain unless we violate this idea of perfect symmetry, they may therefore be regarded as the perfect embodiments of this idea as it exists in the human mind.

But it is well known that partial symmetry is frequently as pleasing as perfect symmetry. For example, in disposing ornaments in a square room, perfect symmetry requires the architect to place the same ornament on each of the four walls. Yet a certain symmetry is preserved, and at the same time variety obtained, if only the ornaments on the opposite walls are made alike; and such a disposition is less stiff than the other.

Now if we apply this principle in arranging our planes around the axes, we shall obtain in addition to the first group of seven forms two other and sim-

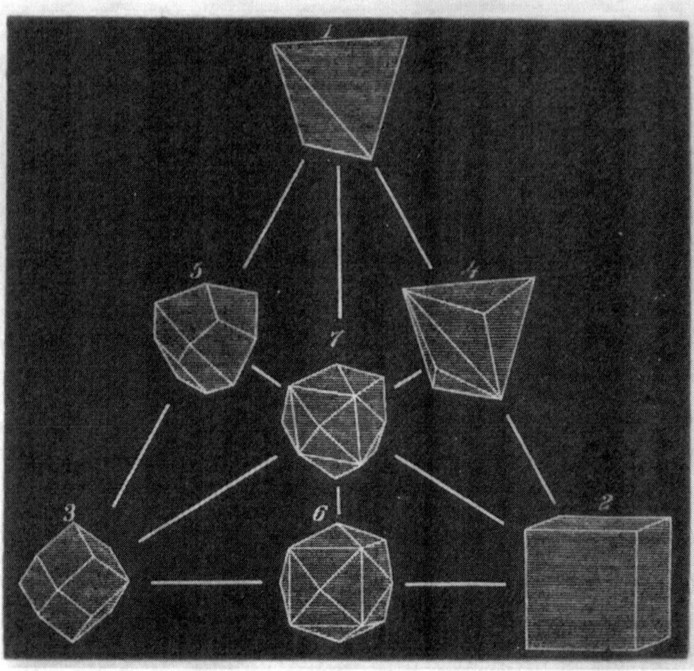

Fig. 13.

ilar groups, which may also be regarded as the embodiment of the idea of symmetry, although not of perfect symmetry. These groups are represented in Figs. 13 and 18, and the forms numbered alike correspond in all three. The manner in which the forms of the *second group* are derived from those of the *first* is illustrated in Figs. 14, 15, 16, and 17.

Thus in Fig. 14 we have first the shaded form of the octahedron No. 1 of the *first group*. Next we have, circumscribing the first, the outlines of the *tetrahedron*, as it is called, which is No. 1 of the *second group*, and it will be noticed that the planes of the tetrahedron correspond precisely in position to the alternate planes of the octahedron; but they are proportionally much larger, and there are only half as many. To carry out a previous illustration, it is as if, instead of hanging a picture on every wall of a square room, we should hang one on only every other wall.

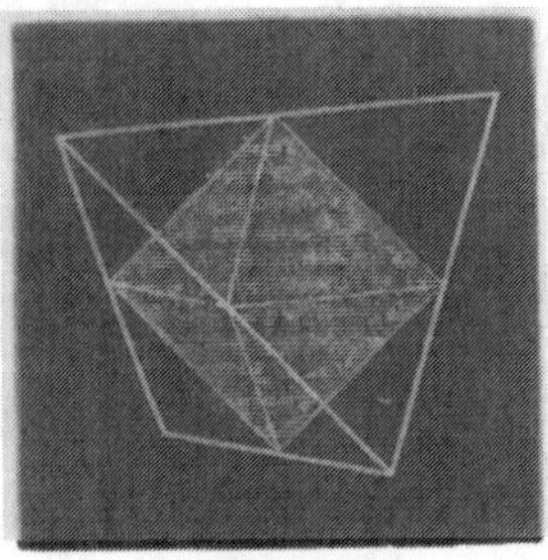

Fig. 14.

If now you will turn back to No. 4 of the *first group*, it will be seen that, corresponding to each plane of the octahedron, there is in this form a set of three planes;—just as if, instead of hanging a single picture on each wall of a square room, we should hang in each place a group of three pictures. Returning again to No. 4 of the *second group*, it will be evident from inspection, that in this form a set of three planes occupies the position of each plane of the tetrahedron, — that is to say, the groups of three pictures have been hung on only every other wall; and if next you examine Fig. 15 — where the form of the *second group* is circumscribed around the corresponding form of the *first group* — you will readily perceive that the

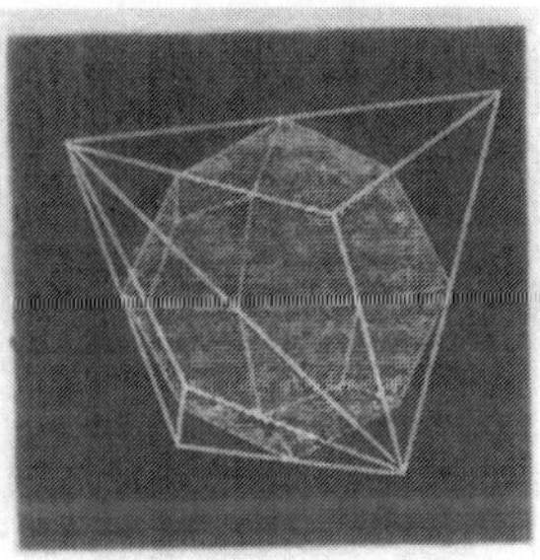

Fig. 15.

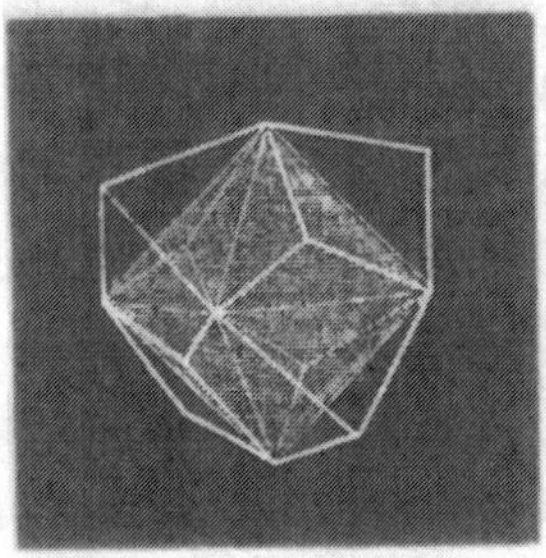

Fig. 16.

sets of three planes on the outer form occupy the same position as the alternate sets of three planes on the inner form.

In like manner, Fig. 16 shows the relation between the two forms numbered 5 in each group, and Fig. 17, that between the two forms numbered 7. In No. 5 of the *first group*, as in No. 4, there are three planes in the place of every plane of the octahedron;—that is, three pictures have been hung on each wall, but they have been differently arranged among themselves. In No. 5 of the *second group* the relative arrangement of the three planes is the same as before, but the sets are placed only in every other position,—that is, the pictures are grouped in the new way, but hung only on every other wall. So also in No. 7 of the *first group* we have a set of six planes in the place of every plane of the octahedron,—six pictures peculiarly grouped and hung on every wall. In No. 7 of the second group we have the same set of six planes placed only in every other position,—six pictures grouped as before and hung on every other wall.

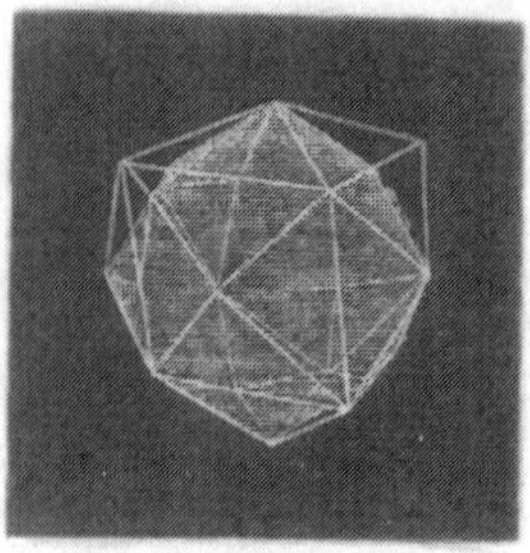

Fig. 17.

The idea of symmetry which is carried out in all these forms is evidently that of the octahedron, which corresponds to the square room in our illustration.

Now if, in developing this same idea, we attempt

to derive a hemihedral* from either of the forms No. 2, 3, or 6 of the *first group,* we shall be surprised to find that no new form results. The reason is this. If you divide up the planes of either of these forms into sets which shall correspond in position to the faces of the octahedron, it will be found that the same plane always belongs to two or more sets. It makes, therefore, no difference whether we place the sets in every position, or only in every other position. The same form results in both cases. It is as if the ornaments in our square room were brackets, which we could not put in their required places in the corners against one wall, without having them touch the adjoining wall also.

Thus it is that these three forms, regarded from one point of view, are *holohedral,*† and belong to the *first group,* while, regarded from another, they are *hemihedral,** and belong to the *second.* Such anomalies as these are not uncommon in science. It is frequently the case that, to our finite minds at least, the same substance, or the same form, is an embodiment of two different ideas, so that, while, regarded from one point of view, it falls into one group of our classification, regarded from another, it belongs to quite a different one.

Passing lastly to the *third group,* Fig. 18, we find another set of hemihedral forms, but they are based on quite a different idea of symmetry; not the idea embodied in the octahedron, but that which is real-

* Having half the number of possible planes.

† Having the full number of possible planes.

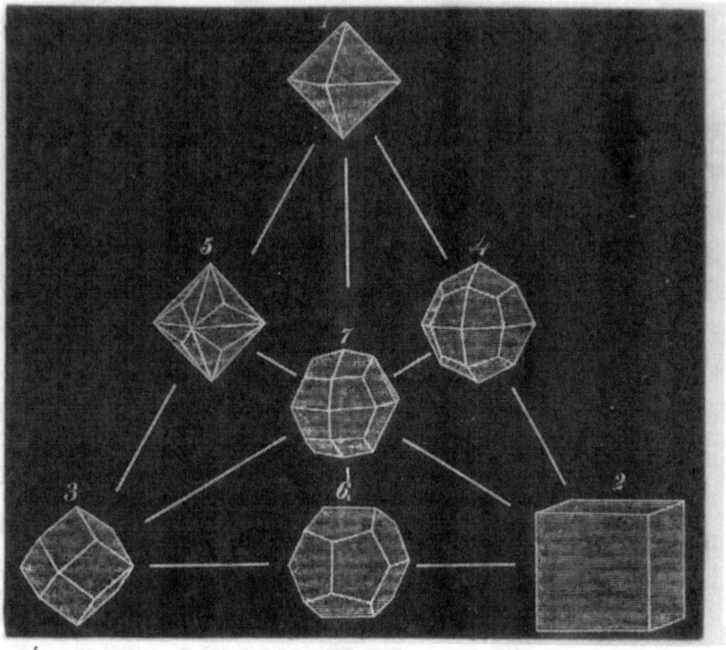

Fig. 18.

ized in the form No. 6 of Fig. 12, called the tetrakis-hexahedron (from Greek words meaning four times six planes). This form is bounded by twenty-four planes placed in the positions already indicated in Fig. 10. If now we place planes only in every other one of these twenty-four positions — as shown in Fig. 19 — we shall obtain the form No. 6 of Fig. 18, bounded by twelve pentagons.

Comparing next No. 7 of Fig. 12 with No. 6 of the same group, we shall find that a pair of planes in the first corresponds to each single plane in the last. When, therefore, we place a pair of planes in every one of the twenty-four positions occupied by the planes of No. 6, we get the form No. 7, bounded by its forty-eight planes; but if we place a pair of planes only in every other position, as shown by Fig. 20, we shall develop the form represented by No. 7 of Fig. 18, which is bounded by only twenty-four planes. Applying now the same idea to the other five forms of the *first group*, we find that no new forms are produced; so that of the forms of the *third group* only two differ from those of the *first*, and the reason here is the same as that already given to explain the similar cases in the *second group*. The like forms may be regarded as embodying at the same time two different ideas of symmetry.

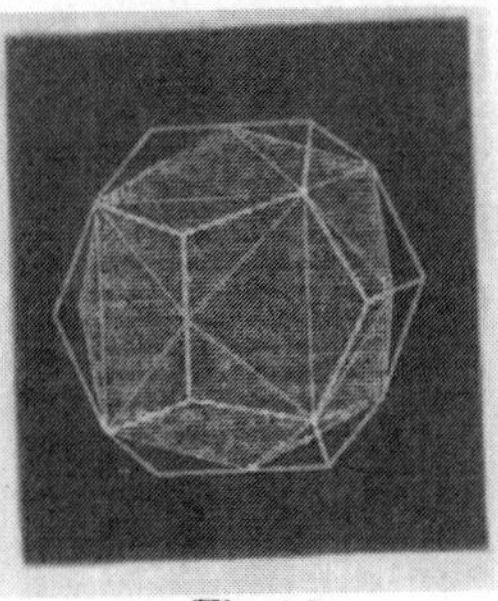

Fig. 19.

I could easily show that the hemihedral forms in each of the last two groups are related among themselves precisely in the same way as are the holohedral forms of the first group, but I am afraid I have already been too abstruse, and I will not further trespass on your patience. Suffice it then to say, that the three groups of forms we have figured are the only forms which we can obtain by arranging planes around the monometric system of axes, according to the idea of symmetry as it exists in our minds, and they may therefore be regarded as the complete embodiment of that idea.

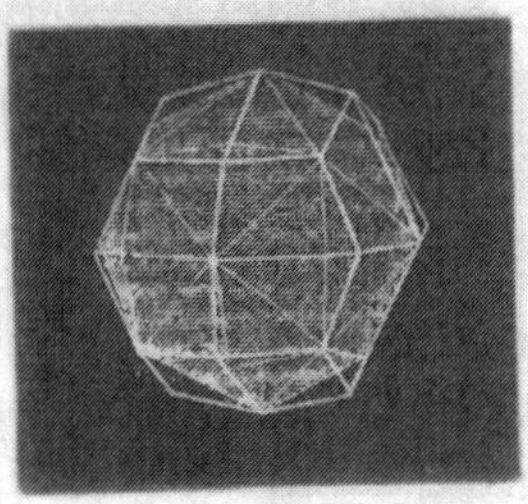

Fig. 20.

Let us now turn to Nature, and compare her handiwork with these products of our own thought. Mineralogists have found that the crystals of gold, silver, copper, the diamond, the ruby, common salt, alum, fluor-spar, galena, iron pyrites, and a very large number of other substances, are all formed around the monometric system of axes, and these crystals have invariably* one or the other of the forms we have described. Moreover, the forms of the crystals of each substance are limited to those of one group.† Thus we find crystals of fluor-spar in all the forms of the *first* group, but never in those which are peculiar to the other two. In like manner, the crystals of a complex ore of copper, called tetrahedrite, have exclusively the forms of the second group, while those of iron pyrites occur only in the forms of the third. Each substance follows a certain idea of symmetry, which seems to be inherent and a part of its very nature; so much so, that when from any cause the character of the symmetry changes, the substance loses its identity, and even if its chemical composition does not change, it becomes to all intents and purposes a different substance.

What has been shown to be true of the regular system of axes is equally true of the five other systems which have been observed in nature. By arranging planes symmetrically around these axes we

* In some cases a substance crystallizes in the forms of two different systems, but then the properties differ in the two states as widely as the form. The diamond and plumbago are examples of this.

† There may be exceptions to this rule, but if so they have not been distinctly stated.

can develop an immense number of forms by a pure process of thought, and it is these forms alone which we find among natural crystals. The products of Nature's laboratory correspond, then, exactly to the results of our own thoughts; and how can we resist the conclusion that they are the manifestations of the thoughts of an intelligent Creator? In the language of science, the crystal is said to obey the law of symmetry; but obviously this law is merely the reflection of the same simple idea which exists in our own minds, and which must have previously existed in the mind of God. The whole science of crystallography is a development of this idea of symmetry. Like geometry, it is a product of pure thought, and its truths are entirely independent of their material forms. Indeed, the mineral kingdom, so far as it is known, does not perfectly represent the idea of symmetry, even as it exists in the human mind. There are possible forms which have never been obtained in nature, and the science, even as we know it, could never have been developed by observation alone.

By following out the simple idea of symmetry, which is common to all men, we have found that the results of our own thought perfectly agree with the facts of nature. Let us now take another of the primary ideas which exist in the human mind, and see how fully that is realized in the material creation. The idea of number is as inherent in the mind as that of symmetry. I shall not attempt to discuss its origin or trace its devel-

Idea of number.

opment; but assuming, as all will admit, that the results of human skill constantly exhibit simple numerical relations, let us inquire whether the same characteristic may not be discovered in nature.

I have already defined a chemical element as a substance which has never as yet been decomposed,

Chemical equivalents.

and all the matter with which man is acquainted is composed of one or more of sixty-two elementary substances. When two of these elements unite together to form a compound body, the proportions in which they combine are not decided by chance. You cannot unite these elementary substances in any proportion you please. The proportion in each case is determined by an unvarying law, and the amounts required of either substance are weighed out by Nature in her delicate scales with a nicety which no art can attain. Thus, for example, 23 ounces of sodium will exactly unite with 35.5 ounces of chlorine; and if you use precisely these proportions of the two elements, the whole of each will disappear and become merged in the compound, which is our common table salt. But if, in attempting to make salt, we bring together clumsily 23.5 ounces of sodium and 35.5 ounces of chlorine, Nature will simply put the extra half-ounce of sodium on one side, and the rest will unite.

The numbers in the table on page 290 give the proportions by weight in which the elements unite. For example, eight pounds of oxygen combine with six pounds of carbon, with fourteen pounds of nitrogen, and so with the numbers given for the other substances. We call these numbers combining proportions, and sometimes chemical equivalents, because in many cases they represent equivalent quantities of the respective elements. Thus, if we remove sulphur from a chemical compound, and fill its place with oxygen, it is found that for every sixteen ounces of sulphur removed we must supply eight ounces of oxygen. Hence, in a chemical sense, eight ounces of oxygen are the equivalent of sixteen ounces of sulphur; but the same is not necessarily true of all the elements.

The standard of these weights is of course arbitrary, but if one number stands for pounds, all the rest stand for pounds, or if one stands for ounces, all the rest stand for ounces. It is usual, however, to leave the standard indefinite, and speak of so many parts. Again, the weights have only relative values; but if we give to any one a definite value, all the rest assume definite values. Our unit must necessarily be more or less arbitrary. Most chemists take hydrogen for the unit of weight, and the numbers given in the table express the equivalent weights of the other elements calculated on this assumption. But we might take any one of the elements as our starting-point, and many of the European chemists use a system of equivalents calculated on the assumption that the equivalent of oxygen is 100. This assumption gives an entirely different system of numbers; but the difference is of no importance so long as the relative values remain unchanged.

TABLE OF ELEMENTARY SUBSTANCES, WITH THEIR EQUIVALENTS OR COMBINING PROPORTIONS AND SYMBOLS.

Element.	Symbol.	Equivalent.	Element.	Symbol.	Equivalent.	Element.	Symbol.	Equivalent.
		H = 1			H = 1			H = 1
Aluminum	Al	13.75	Hydrogen	H	1.0	Rubidium	Rb	85.36
Antimony (Stibium)	Sb	122.0	Iodine	I	127.0	Ruthenium	Ru	52.11
Arsenic	As	75.0	Iridium	Ir	98.56	Selenium	Se	39.75
Barium	Ba	68.5	Iron (Ferrum)	Fe	28.0	Silicon	Si	14.0
Bismuth	Bi	210.0	*Lanthanum*	La	46.0	Silver (Argentum)	Ag	108.0
Boron	B	10.9	Lead (Plumbum)	Pb	103.5	Sodium (Natrium)	Na	23.0
Bromine	Br	80.0	Lithium	L	7.0	Strontium	Sr	43.75
Cadmium	Cd	56.0	Magnesium	Mg	12.16	Sulphur	S	16.0
Calcium	Ca	20.0	Manganese	Mn	27.5	Tantalum	Ta	68.8
Carbon	C	6.0	Mercury (Hydrargyrum)	Hg	100.0	Tellurium	Te	64.5
Cerium	Ce	46.0	Molybdenum	Mo	48.0	*Terbium*	Tb	
Chlorine	Cl	35.5	Nickel	Ni	29.5	*Thallium*	Tl	204.0
Chromium	Cr	26.27	Niobium (Columbium)	Nb	48.8	*Thorinum*	Th	59.5
Cobalt	Co	29.5	Nitrogen	N	14.0	Tin (Stannum)	Sn	59.0
Cœsium	Cs	133.0	Osmium	Os	99.5	Titanium	Ti	25.0
Copper (Cuprum)	Cu	31.75	Oxygen	O	8.0	Tungsten (Wolfram)	W	92.0
Didymium	D	48.0	Palladium	Pd	53.24	Uranium	U	60.0
Erbium	E		Phosphorus	P	31.0	Vanadium	V	68.46
Fluorine	F	19.0	Platinum	Pt	98.56	*Yttrium*	Y	
Glucinum	G	4.66	Potassium (Kalium)	K	39.0	Zinc	Zn	32.75
Gold (Aurum)	Au	196.66	Rhodium	Ro	52.16	*Zirconium*	Zr	44.75

NINE SERIES.

Elements. 8 + n 9.	Equivalents. Theo.	Observ.	n	Usual Condition.	Specific Gravity.	Boiling Point.	Sp. Gr. as Gas.	Color of Gas.	Affinity for Oxygen.	Affinity for Hydrogen.
Ozone	8	8	0	Gas			1.11	Colorless		Intense
Fluorine	17	19	1	Gas					No compound known	Less intense
Cyanogen	26	26	2	Gas			1.81	Colorless	Very feeble	Very strong
Chlorine	35	35.5	3	Gas	1.33	—58° nearly	2.47	Yellow	Very feeble	Very strong
Bromine	80	80	8	Liquid	3	145°	5.54	Red	Feeble	Less strong
Iodine	125	127	13	Solid	5	347°	8.72	Violet	Decided	Feeble

EIGHT SERIES.

Elements. 8 + n 8.	Equivalents. Theo.	Observ.	n	Usual Condition.	Specific Gravity.	General Appearance.	Melting Point.	Boiling Point.	Sp. Gr. as Gas.
Oxygen	8	8	0	Permanent Gas		Colorless			1.11
Sulphur	16	16	1	Solid	2	Yellow	230°	824°	2.23
Selenium	40	39.7	4	Solid	4.3	Reddish Brown	392°	Red heat	5.68
Tellurium	64	64.5	7	Solid	6.1	Metallic	661°	White heat	9.00

SIX SERIES.

Elements. 8 + n 6.	Equivalents. Theo.	Observ.	n	Usual Condition.	Specific Gravity.	General Appearance.	Melting Point.	Boiling Point.	Sp. Gr. as Gas.
Nitrogen	14	14	1	Gas		Colorless			0.97
Phosphorus	32	31	4	Solid	1.77	Transparent Solid	111°	554°	4.42
Arsenic	74	75	11	Solid	5.67	Sub-metallic	Low red heat	Red heat	10.20
Antimony	122	122.3	19	Solid	6.72	Metallic	840°	Volatile at white heat	
Bismuth	212	210	34	Solid	9.82	Metallic	513°	Barely volatile	

Dr. Prout was the first to notice that many of the chemical equivalents were simple multiples of the equivalent of hydrogen, and he thought that, if this equivalent were taken as unity, the other equivalents could all be expressed by whole numbers. The progress of chemistry has not tended to confirm this view, but on examining the table it will be noticed that a large number, at least, of the chemical equivalents on the hydrogen scale are whole numbers, and this is especially true of those elements which have already been mentioned as forming almost the whole of the earth's crust. Thus we find in these elementary forms of matter — the blocks with which the universe has been built — the same simple numerical relations which everywhere appear in the constructions of man.

Many of the equivalents are whole numbers.

Still simpler numerical relations will be discovered if we consider the different compounds formed by the same elements. The proportions of the elements which enter into such compounds always bear a very simple numerical relation to each other. Thus there are five compounds of oxygen and nitrogen which contain these elements in the proportions indicated in the following table.

Law of multiple proportions.

Compounds of Oxygen and Nitrogen.

	Nitrogen.	Oxygen.
Protoxide of Nitrogen	14 parts.	8 parts.
Deutoxide of Nitrogen	14 "	$8 \times 2 = 16$ "
Nitrous Acid	14 "	$8 \times 3 = 24$ "
Hyponitric Acid	14 "	$8 \times 4 = 32$ "
Nitric Acid	14 "	$8 \times 5 = 40$ "

It will be noticed that the proportions of oxygen in these compounds are in all cases simple multiples of eight, the proportion in the first. In like manner, the compounds of manganese with oxygen show similar relations.

Compounds of Oxygen and Manganese.

	Manganese.	Oxygen.
Protoxide of Manganese	27.5 parts.	8 parts.
Sesquioxide of Manganese	27.5 "	$8 \times 1\frac{1}{2} = 12$ "
Hyperoxide of Manganese	27.5 "	$8 \times 2 = 16$ "
Manganic Acid	27.5 "	$8 \times 3 = 24$ "
Permanganic Acid	27.5 "	$8 \times 3\frac{1}{2} = 28$ "

The relation is not quite so simple as in the other case, but still the same general truth is evident, and these two examples are fair illustrations of what has been observed throughout the whole range of chemical compounds.

Combining proportions by volume.

A large number of the chemical elements are either found naturally in the aeriform condition, or else can be converted into gases by a moderately high temperature, and some remarkably simple numerical relations will appear if we consider the proportions in which these gases combine by volume. We gave above a table of combining proportions by weight, and we give on the next page a table of combining proportions by volume.

What could be simpler than the numerical relations presented by this table? These gases combine either in the proportion of one volume

Combining Proportions by Volume.

Name of Gas.	Volume.	Sp. Gr. Calculated.*	Sp. Gr. Observed.
Arsenic	1	10.365	10.65
Brominic	2	5.526	5.54
Chlorine	2	2.450	2.47
Cyanogen	2	1.797	1.81
Fluorine	2	1.301	
Hydrogen	2	0.069	0.069
Iodine	2	8.768	8.716
Mercury	2	6.976	6.976
Nitrogen	2	0.968	0.971
Oxygen	1	1.106	1.106
Phosphorus	1	4.334	4.42
Selenium	1	4.406	5.68
Sulphur	1	2.211	2.23

(that is, one quart, one pint, one cubic foot, or whatsoever unit of volume we employ) of one gas, to one of the other, or else one volume of one to two of the other. It is true, however, that combination may take place in any simple multiples of these proportions, as in the case of the equivalents by weight. For example, the five compounds of oxygen and nitrogen contain respectively,—

	Oxygen.	Nitrogen.
Protoxide of Nitrogen	2 volumes.	1 volume.
Deutoxide of Nitrogen	2 "	2 "
Nitrous Acid	2 "	3 "
Hyponitric Acid	2 "	4 "
Nitric Acid	2 "	5 "

* These specific gravities are calculated on the assumption that the combining proportions by volume are exact. It will be noticed that they very closely agree with the observed specific gravities given in the last column.

But there are numerical relations between the chemical elements even more remarkable than those we have studied. These substances, which we call elements solely because they have never been decomposed, vary very greatly in their properties. At the ordinary temperature of the air, some are gases, some are liquids, and some are solids. Some are highly corrosive, like chlorine, and others perfectly inert, like nitrogen. Some have a bright metallic lustre, like gold or silver, and others are dull, like charcoal or sulphur. Indeed, we find among these substances almost every possible variety of qualities.

Classification of elements.

These elements are not to be regarded, however, as so many distinct creations, each endowed with the peculiar properties which fit it for the uses it subserves in the universe, but each entirely independent of all the rest. On the contrary, the whole world of matter is but one creative thought, expressed with various modifications in the different elements, but yet one thought comprehending all. It is true that we have not obtained more than the most distant glimpses of this Divine plan, but still we have seen enough to convince us that there is such a plan underlying the whole material creation.

Of the sixty-four substances which have never been decomposed, and which are therefore regarded as elements, a large number are so exceedingly rare and so difficult to obtain that their properties have not been fully studied, and their relations to one another are not yet known. Disregarding those italicized in the table on page 290, the remaining elements may be classified in six families or series, three of which are given on the opposite page of the same table. We shall best be able to understand the principles of the classification by examining in detail one of the families of elements. We might take either of the six, and I have selected the Nine Series only because the members of this family of elements are probably more familiar to you than any others. The elements which compose the nine series are all highly active and corrosive chemical agents. They all form analogous compounds. They all have, so far as is known, the same crystalline form, and may replace one another in a chemical compound without changing the form of the substance. Moreover, they resemble one another in many other properties which we have not time to discuss in this connection.

If the resemblances are great, however, the differences are great also; but then these differences follow a simple and well-established law of progression. The properties of this family of elements vary, as we descend the series, by regular steps. Let me illustrate. Consider first the physical condition of these substances. Ozone is the active modification of oxygen, and so persistently does it retain its aeriform condition that no amount of mechanical pressure and cold which we can command is sufficient to reduce it to a liquid. The next element, fluorine, is also a gas, but it is so

Law of progression.

highly corrosive that we have not yet been able to experiment upon it. Cyanogen is a very remarkable substance. It is really a compound, but its properties resemble so closely those of the other elements of the series that I have classified it with them. It is a gas which can by great pressure and cold be changed to a liquid. Chlorine, although generally known as a gas, can be liquefied still more readily. Bromine is a very volatile liquid, which, at the temperature of 145° Fahrenheit, becomes a gas. Iodine, finally, is a solid, but a solid which is very fusible, and becomes a gas at a temperature of 347°. At the temperature of a common bread-oven, all these substances would be gases, and, as is shown in the eighth column of the table, the densities of these gases gradually increase as you descend in the series. Pass now to the color of these elements when in the state of gas, and you will find that there is a beautiful chromatic gradation between them. Ozone is colorless; chlorine is yellow, inclining to green; bromine, orange-red; and iodine, violet. Consider next the equivalents of these elements, and you will discover that they stand in a very simple relation to each other. The equivalent of ozone is 8; that of fluorine is approximately $8 + 9 = 17$; that of cyanogen, $17 + 9 = 26$; that of chlorine, $26 + 9 = 35$. It is probable that, if the series were complete, the equivalent of the next element to chlorine would be $35 + 9 = 44$, and those of the three succeeding elements, 53, 62, and 71 respectively; but no such elements are known. The next step, however, gives us eighty, which is the equivalent of bromine. Then again there are four steps not represented, but at the fifth step we come to iodine, whose equivalent is 125. You notice that the common difference of these numbers is in every case nine, and for this reason I have named this family of elements the Nine Series.

A relationship like that I have just described, exists among the members of the other five families into which I have classified the chemical elements. The properties of the elementary substances in each family vary according to a similar law of progression; but the number which expresses the common difference of the equivalents is different for each series, and characterizes the family. The numbers besides 9 are 8, 6, 5, 4, and 3, respectively, and in each case I have given to the series the name of the numeral which distinguishes it.

How evidently, then, the properties of the elements conform to the idea of number! It is true that our classification is at fault in many particulars, and I do not suppose that the order of progression here given is more than a very distant approximation to the actual law of nature; but the fact that some such law exists, and that it is based on simple numerical relations, I believe is beyond a doubt. The same principle of classification applies to compound bodies as well as to elements, and I confidently look forward to the time when it will be possible to classify all substances in one comprehensive scheme, and to represent the relations of their properties by simple arithmetical formulæ.

Intervals between the orbits of the planets.

Similar numerical relations are found throughout the whole universe of matter. In the solar system, for example, with the exception of Neptune, the intervals between the orbit of Mercury and the orbits of the other planets go on doubling, or nearly so, as we recede from the Sun. Thus the interval between the Earth and Mercury is nearly twice as great as that between Venus and Mercury, the interval between Mars and Mercury nearly twice as great as that between the Earth and Mercury, and so on. Again, if we compare the periods of revolution around the Sun, expressed in days, we shall find another simple numerical relation, as shown by the following table.

Law of Periodic Times.

	Observed.	Theoretical.		Fractions.
Neptune	60,129	62,000		
Uranus	30,687	31,000		$\frac{1}{2}$
Saturn	10,759	10,333		$\frac{1}{3}$
Jupiter	4,333	4,133		$\frac{2}{5}$
Asteroids	1,200 to 2,000	1,550		$\frac{3}{8}$
Mars	687	596		$\frac{5}{13}$
Earth	365	366	$\frac{8}{13}$	$\frac{8}{21}$
Venus	225	227	$\frac{13}{21}$	
Mercury	88	87		$\frac{13}{34}$

It will be noticed that the period of Uranus is $\frac{1}{2}$ that of Neptune, the period of Saturn $\frac{1}{3}$ that of Uranus, the period of Jupiter about $\frac{2}{5}$ that of

Saturn, the period of the Asteroids about $\frac{3}{8}$ that of Jupiter, the period of Mars about $\frac{5}{13}$ that of the Asteroids, the period of Venus about $\frac{8}{21}$ that of Mars, and the period of Mercury about $\frac{13}{34}$ that of Venus. The successive fractions are very simply related to each other, as will at once appear on writing them in a series,

Ratios between the periods of the planets.

$$\frac{1}{2}, \frac{1}{3}, \frac{2}{5}, \frac{3}{8}, \frac{5}{13}, \frac{8}{21}, \frac{13}{34}, \text{\&c.}$$

Notice that, after the first two, each succeeding fraction is obtained by adding together the numerators of the two preceding fractions for a new numerator, and the denominators for a new denominator. From this series, however, the Earth is excluded. Its time of revolution is almost exactly $\frac{8}{13}$ of that of Mars, and that of Venus nearly $\frac{13}{21}$ of that of the Earth; but although these fractions do not fall into the above series, they are members of a complementary series beginning

$$\frac{1}{2}, \frac{2}{3}, \frac{3}{5}, \frac{5}{8}, \frac{8}{13}, \frac{13}{21}, \text{\&c.}$$

This simple relation was discovered by Professor Peirce, and he has proposed an explanation for the anomaly presented by the Earth. But it is not important to dwell on this point. Our only object was to show that simple numerical relations appear in the planetary system, and this, as I trust, has been fully illustrated.

Passing now to the vegetable kingdom, we find again the same numerical laws. The leaves of a plant are always arranged in spirals around the stem. If we start from any

Arrangement of leaves.

one leaf, and count the number of leaves around the stalk and the number of turns of the spiral until we come to a second leaf immediately over the first, we find that for any given plant, as an apple-tree for example (see Figs. 21 and 22 *), the number

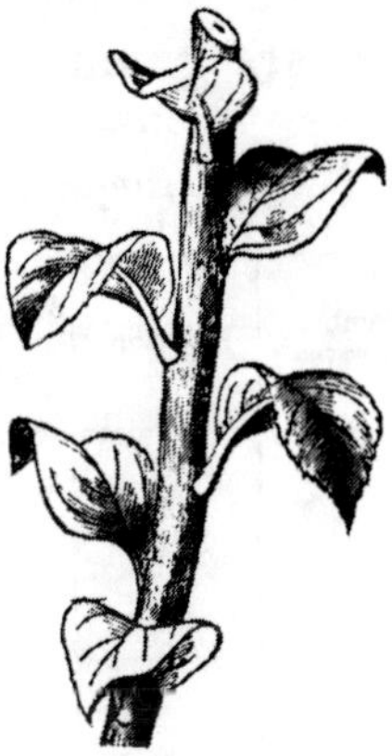

Fig. 21.

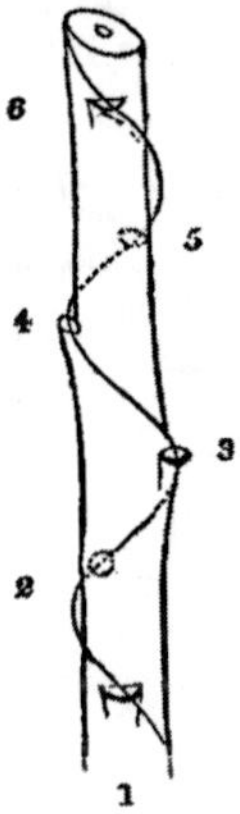

Fig. 22.

of leaves and the number of turns of the spiral are always absolutely the same. The simplest arrangement is where the coincidence occurs at the second leaf, after a single turn of the spiral; and this may be expressed by the fraction $\frac{1}{2}$, whose numerator denotes the number of turns of the spiral, and whose denominator the number of leaves. The next simplest arrangement is when the coincidence occurs at the third leaf, after a single turn of the spiral, and may be expressed by the fraction $\frac{1}{3}$. These two fractions express respectively the greatest and the smallest divergence between two successive leaves which has been observed. The angle between two

* These figures are from Gray's Botanical Text-Book.

successive leaves, therefore, is never greater than 180°, or half the circumference of the stem, and never less than 120°, or one third of the circumference. The arrangement next in simplicity is where the coincidence occurs at the fifth leaf, after two turns of the spiral, as is represented in the preceding figures. Other examples are given in the table

LAW OF PHYLLOTAXIS (*Leaf-Arrangement*).

Name of Plant.	Number of Turns of Spiral.*	Number of Leaves.*	Fraction.	Angle of Divergence. between two successive Leaves.
Grasses,	1	2	$\frac{1}{2}$	180°
Sedges,	1	3	$\frac{1}{3}$	120°
Apple, Cherry, Poplar,	2	5	$\frac{2}{5}$	144°
Holly, Callistemon, Aconite,	3	8	$\frac{3}{8}$	135°
Rosettes of the Houseleek, Cones of the White Pine,	5	13	$\frac{5}{13}$	138° 28′
Cones of the European Larch,	8	21	$\frac{8}{21}$	137° 9′
Certain Pine Cones,	13	34	$\frac{13}{34}$	137° 39′
Certain Pine Cones,	21	55	$\frac{21}{55}$	137° 27′
Typical arrangement which would expose to the Sun's rays the greatest leaf-surface,				137° 30′ 28″

which follows, and it will be seen that we have precisely the same series of fractions in the arrangement of leaves around the stem of a plant which appears in the periods of the planets. The fractions of this series are all gradual approximations to a

* Before a leaf occurs vertically over the first.

mean fraction between $\frac{1}{2}$ and $\frac{1}{3}$, which would give the most nearly uniform distribution possible to the leaves, and expose the greatest surface to the sun.

But this law does not stop with the plants. The same series of fractions expresses also the spiral arrangement of the tentacles of the Polyp and of the spines of the Echinus. Thus through the whole realm of nature, from the structure of the crystals to the dimensions of the human form, a similar numerical simplicity is preserved.

Have you never recognized the composition of your friend in some anonymous literary article, by a peculiar phraseology, a turn of style, or a method of thought which no artifice can conceal? Have you never felt a glow of pleasure when you unexpectedly discovered on the walls of a picture-gallery the work of a well-known artist, marked by some peculiarity of grouping or coloring? Has your attention never been quickened when an orchestra has suddenly struck into a new theme of a favorite composer, never heard before, but unquestionably his? If you have experienced these or similar emotions, you know something of the force with which such numerical laws impress the mind of the student of nature, and you also know how difficult it is to make the power of such impressions understood. I wish I could give you a full conception of this power; for you cannot otherwise feel the full force of the evidence which these facts afford. They point directly to an intelligence in nature like our own, and they are a seal to the declaration of the Bible, that man was created in the image of his God.

The broken porticos of the Parthenon still stand on the Acropolis at Athens to excite the imitation and win the admiration of the architect. That beauty of outline and those faultless proportions, which modern art has copied, but never excelled, all depend on an exact conformity of all the parts to the laws of symmetry and to simple numerical ratios. We justly regard that ruined temple as the evidence of the highest intelligence; and when we find the same symmetry, the same numerical ratios, appearing everywhere in nature, how can we refuse to admit that they also are the evidence of intelligence and thought? Moreover, since the laws of symmetry and number pervade the whole universe, from the structure of the solar system down to the organization of a worm, they prove, if they prove anything, that the whole is the manifestation of the thoughts of the one great Jehovah, who in the beginning created all things by the word of His power.

LECTURE X.

ARGUMENT FROM GENERAL PLAN (CONCLUDED).—NECESSARY LIMITATIONS OF SCIENTIFIC AND RELIGIOUS THOUGHT.

In my last Lecture I endeavored to show that the laws of nature, so far from proving that the world is governed by an inexorable necessity, furnish the strongest evidence of an overruling mind. We must be careful, however, not to misinterpret this evidence; for analogies like those we have studied led Schelling and the philosophers of his school to regard outward nature not merely as the result of the Divine Thought, but as identical with that thought, and inseparable from it. Indeed, there are many among us who regard the material universe as the manifestation of God, in the same intimate sense in which our bodies are the manifestation of our own personality; who therefore believe that the world is and always has been a part of His Eternal Being, and who look upon the laws of nature not merely as the manifestation of an Infinite Intelligence, but as a part of that Intelligence itself.

The error of pantheism.

This philosophy may be made to appear very attractive, and even very reverential; but when followed out to its logical consequences, it reduces God to the level of nature, and merges His being in the matter He created. We must be as careful to avoid the snares of pantheism, as the slough of materialism. Both are equally destructive of true religion, and, although they lie on opposite sides of the Christian's path, they lead to the same result; and if once enticed from the narrow way, the Christian will be fortunate if Faith rescues him from the peril before he plunges utterly into the gulf of atheism. We must not confound the Creator with the creature. There is a personal God above all and over all, and although nature manifests His intelligence, its material forms are only the reflection, not the substance, of His Being. The error of the pantheist arises from a too superficial study of nature, and if we examine more closely the analogies between the laws of nature and the results of human thought, I am confident we shall find that the created forms may be readily distinguished from the Intelligence which gave them being.

Distinction between the conception and the execution of a work.

In every human work, we may always distinguish two things, the conception and the execution, and the last never exactly conforms to the first. For example, in one of the grand Gothic cathedrals of our mother country we see united in the plan, first, the idea of the cross, the emblem of our Christian faith; then the spire, typifying the aspiration of the soul; and lastly, the long aisles, whose pointed arches and delicate tracery have been copied from the interlacing branches of God's first temple. The combination of these ideas may be said to be the conception of the cathedral; but how differently has this conception been embodied in the numerous cathedral churches of England! Besides the peculiar caprices of the architect or builder, we can trace in each church an evident adaptation of the parts to special purposes. Here a "lady chapel" has been included in the design, and here the mausoleum of a king or a prelate; here a portion has been adapted to the reading of the service, and here to the session of the ecclesiastical court; but however varied the execution, the same conception is evident in all. So it is in all architecture. Our modern dwellings are built after a few general types, and the conception is very nearly the same in all houses of any one class. But how differently a skilful architect will arrange the details, and adjust them to the circumstances of the location, to the wants of the family, or the taste of the owner! and no one knows better than he that the conception of the building is one thing, and the execution of that conception a very different thing.

Illustrated by architecture.

In the higher forms of art, the same truth appears even more strikingly. The Transfiguration of Raphael, that masterpiece of painting, does not hold you breathless before it so much by what it actually represents, as by what it embodies and helps you to realize. He who sees merely what is painted on the canvas will turn away disappointed, but in the soul of the true student of art, who enters into the spirit of the great

Illustrated in art.

painter, the conception grows as he gazes, until he becomes transported and gains a vision of the splendor of the Mount. In like manner, it is not that lovely female face which has endeared the Sistine Madonna to so many hearts, and made Dresden one of the shrines of the world. In mere point of execution this picture may be surpassed by many works of living artists; but the conception of a pure mother's love has been nowhere embodied as there, and that is the charm. You stand before the Laocoön until the blood grows cold and the muscles writhe in sympathy, and then you look at the motionless statue and wonder whence comes the power. It is not in the skilfully chiselled marble, but it is in the conception of the unknown artist, which the petrified forms suggest. So it is everywhere with the works of man; the conception can always be distinguished from the embodied fact. But what need of illustration? Who does not know the difference between the two, and who has not sadly experienced how far his best efforts fall short of his ideal? The thought, the conception, how noble! the execution, the reality, how humble!

Turning now to Nature we find the same distinction there between the conception and the facts.

The same distinction in nature.

Nature does not, of course, like man, fall below her ideal for want of power, but she departs from it in order to adapt her works to specific ends. Everywhere the conception, or, as we generally call it, the law, is modified in the execution to insure special adaptation. We may not always be able to discover the adaptation, but in every case we find that the phenomena do not exactly conform to the law. Review for a moment, with this idea, the few examples of natural laws which I have so briefly sketched, beginning with the law of symmetry.

We seldom, if ever, find in nature crystals having that regularity of form or that perfection of outline represented in our figures. Natural crystals are invariably more or less distorted or imperfect. You would hardly believe the crystal represented by Fig. 23 to be an octahedron, and yet it is merely this familiar form somewhat distorted. I admit that the distortion in this case is unusually great; — still a perfect crystal is at best a very rare exception. It is true that in all cases of distortion the relative inclination of the planes is very nearly constant; but even this is liable to some variation. Moreover, there are several forms of crystals represented in these diagrams,* which are never found in their perfection in nature. They are merely shadowed forth, as it were, on other forms, and so partially that the unpractised eye would never detect them. So true is this, that, as I have before stated, the present science of crystallography could never have been developed by

Illustrated by crystals.

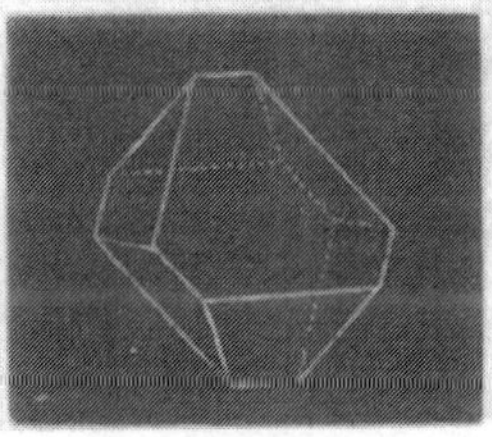

Fig. 23.

* See the three groups of crystals given in the last Lecture.

observation alone. How evident, then, the distinction between the actual crystals and the thought which they embody!

Passing next to chemistry, we notice the same truth in the law of progression by series. We do not find that the properties of the elements in any of the series given in the table on page 291 are exactly those required by the law; but in each case they approximate more or less closely to the ideal type. Thus, if the law were exactly observed, the chemical equivalents would be those given in the column headed Theoretical; and it will be seen that these numbers by no means exactly coincide with the observed equivalents, which for comparison are placed at the side. The same is true of the other properties of these elements.

Illustrated in chemistry.

This simple law of relationship, first observed by myself over ten years since, has been questioned, on the ground that it does not exactly represent the fact; and Dumas, of Paris, while adopting essentially the same system of classification, has modified the details, in order to obtain a more exact conformity, carefully redetermining many of the chemical equivalents, with the evident idea that they must exactly coincide with the law. But surely, if we may reason by analogy, no such precision was to be expected, and, in seeking to realize this illusion, Dumas overlooked the simple conception and substituted a most mechanical and improbable system in its place. It was never supposed that the facts of nature would exactly correspond to the law; for it was known that an exact correspondence between the law and the phenomena is found absolutely nowhere. All I claimed was, that with the various elements there was a general conformity in their properties to this simple conception of series. I supposed that the properties of matter had been distributed according to this general plan; but at the same time, knowing that each element had been specially adapted to a certain specific end, I did not expect to find the conception perfectly realized.

When now we study the elements from this point of view, we at once notice that the departure from the general law is most marked in those substances which are most widely distributed in nature, and where the special adaptations are the most evident. This is especially true of the substances composing the atmosphere. The elements, oxygen, nitrogen, carbon, and hydrogen, and the compounds, water, ammonia, nitric acid, and carbonic acid, are all remarkably unique substances, and were we to examine their external qualities alone, we should never discover their relationship; but on studying carefully their chemical relations, we find that, while they have been adapted with such wonderful skill to the important ends which they have been appointed to subserve, they have also been made to conform, at least in their main features, to the general plan. There is still another point connected with the classification of the elements which is deserving of our notice in this connection. We have already seen that, although sixty-four elements have been discovered, the greater portion of the earth's crust consists of only ten or twelve. Indeed, if the remaining fifty elements were suddenly annihilated, the mass of the globe, so far as we know, would not be sensibly diminished. Indeed, a large number of the elements occur only in some of the rarest minerals, and there in such minute quantities that they can be detected only by the most skilful chemical analysis. That these very rare elements were designed by the Creator to subserve important ends, we need not doubt; but it is certain that they play a very subordinate part on the surface of the globe. For bromine and iodine, and a few others, important applications have been discovered in the arts or in medicine; but the rest, comprising at least one third of all the known elements, have no apparent value except as parts of a general plan. In the light of a utilitarian philosophy they must appear useless; but to the true student of nature they have a significance which transcends everything else. They are parts of a universal order, of a Divine cosmos, which would be incomplete without them. They are the manifestation of Infinite Intelligence. They embody the thoughts of God. Man's material wants and pleasures are not the sole end of creation. In the words of Chevalier Bunsen, "Law is the supreme rule of the universe, and this law is intellect, is reason, whether viewed in the formation of a planetary system or in the organization of a worm."

But we must remember, in discussing this question, that it does not follow, because we cannot discover any important end which these elements subserve on our earth, that they have no practical utility. For after acknowledging the dignity which they acquire when regarded as the characters of that language in which the creative thoughts have been written, and as the appointed means of educating the human race, still it does not seem consistent with that economy of resources which appears in all parts of the Divine plan, that they should have no special functions to discharge in the cosmos. Now I would suggest, but I offer the suggestion in all humility, that these very rare elements may be adapted by their peculiar properties to the thermal conditions of some other planet or some other stellar system. We have seen that those elements which are the most widely distributed over the earth are such as are adapted by their properties to the conditions of organic life on the third planet of the solar system, and it is certainly possible that some different scheme of organic life may be sustained on Mercury or Uranus, in which elements rare to us take the place of oxygen, nitrogen, hydrogen, and carbon, and perhaps also the elements missing in our classification may be found in some other world, revolving around Sirius or Arcturus, where oxygen, sulphur, and iron may be among the rarities of science.

All this is of course the purest hypothesis, and such speculation can lead to no positive results; but the very possibility of the speculation illustrates most pointedly the great truth which I am endeavoring to enforce. The thought which is embodied in the scheme of the chemical elements is something entirely apart from their material forms, and the moment this thought is apprehended by man, it opens to his imagination vistas of possible realities which entirely transcend all human experience.

Illustrated in astronomy.

If next we compare, more carefully than before, the periods of revolution of the planets around the Sun, we shall find that the same principle holds true. The observed periods, you will notice by the table on page 299, do not exactly correspond to the simple ratios which express the law, and the same is true of the distribution of leaves around the stem of a plant, and in fact of all classes of phenomena in nature. In each we observe only a tendency towards a maximum effect, which is the perfect expression of the law, but which is seldom fully reached. The limits of variation are broader in some cases than in others, but we find no case in which the accordance is absolute.

Illustrated in botany and zoölogy.

In none, however, of the purely physical laws is this character so strongly marked as in the structure of animals and plants. It is well known that all organized forms, although so wonderfully diversified, are fashioned after a few general types. In the animal kingdom there are only four general plans, represented by the Radiata, the Mollusca, the Articulata, and the Vertebrata, and all the animals of any one of these great divisions are organized alike. For example, in all vertebrate animals we find essentially the same parts, and frequently where there is no use for a given organ it is still present in a rudimentary condition. Professor Wyman has found rudimentary eyes in the so-called eyeless fish of the Mammoth Cave, and any anatomist will point out to you in the skeleton of a fish, of a reptile, of a bird, or of a quadruped, the bones which correspond to the various parts in the skeleton of a man. In the wing of a bat the bones of the human arm may readily be traced.

Here then is a most obvious distinction between the conception and the execution, and the general plan of the skeleton is preserved even where there is no use for certain parts, and where we might perhaps conceive of a simpler arrangement without them. But, more than this, we find that the variations from what we may regard as the typical form have been obviously made in order to adapt the organs to certain specific ends. The same plan which, developed in its full perfection, appears in the human hand and arm, reappears, more or less fully carried out, in the fore legs of a horse, in the wings of an eagle, and in the pectoral fins of a dolphin; and in each case the organ has been obviously adapted to some special purpose. Special adaptation has thus been most beautifully harmonized with general law, and the conception has been varied in the execution in order to secure some wise and important end. So also it may be that, when in the laws of inorganic nature we cannot discover the cause of a variation, a similar wise purpose is concealed.

This subject is capable of almost indefinite illustration, and the vegetable kingdom is as rich in examples of the principle we have been discussing as the animal. I have not, however, time for further details. The whole ground has been most carefully surveyed by McCosh and Dickie in their excellent work entitled "Typical Forms and Special Ends in Creation," and to this I would refer those who may be interested to pursue the study of these singular facts. Sufficient, I trust, has already been said to show that the phenomena of nature and the results of human thought resemble each other in their very incompleteness.

Nature points to a personal God.

While, therefore, a more careful study has tended to confirm the result at which we arrived in the last Lecture, and has strengthened the impression that the universe was created by an intelligence like our own, we have also found that the analogies of nature point with equal distinctness to the conclusion that this intelligence is a being entirely apart from and infinitely superior to the matter he created or the laws he ordained. If these analogies are worth anything, they point not to a supreme spirit of the universe, pervading and energizing matter, but they prove the existence of a *personal* God; one who can sustain to us the relations of Father, Saviour, and Sanctifier; one whom we can love, worship, and adore.

But it may be urged that I have drawn my illustrations wholly from the phenomenal laws of nature, and entirely overlooked the great dynamical laws, which, like the law of gravitation, are more precise. Moreover, it will be said that the history of astronomy gives us every reason to believe that these very variations to which I have given such importance are merely necessary consequences of some higher law not yet discovered, just as the perturbations of the planetary orbits are the legitimate results of the very law they seem at first to invalidate. I have no doubt that in part, at least, this will be found to be the case. But even in regard to the law of gravitation, there always have been residual phenomena, unexplained by the law, and so probably there always will be, until, as we go on widening our generalizations, the last generalization of all brings us into the presence of that First Cause through whom and by whom all things are sustained.

Force of the argument.

I trust that the striking analogies between the phenomena of nature and the results of human thought, which I have been able so imperfectly to illustrate, have impressed you, as they impress me, with the profound conviction that the order of nature is the manifestation of an Infinite Intelligence, but of an Intelligence apart from, and superior to, the order which it once created and now upholds. If I have failed in my object, it is because I have been unable to bring home these analogies to your understanding. The resemblances are so striking, that I do not believe a mind which is conversant with the facts, and unbiassed by the prejudices of philosophy or of education, can resist the conclusion that this scheme of nature is the manifestation of an intelligence like our own, at least so far as the Infinite can be said to resemble the finite. Men may reasonably entertain differences of opinion in regard to the mode of action of that Being who has created the universe. They may believe that a certain amount of power, together with the germ of all future existence, was implanted in the original chaos, and that the Deity has never interfered with the natural action and the unfolding of the causes which He has thus ordained; but whatever theories of cosmogony may be entertained, short of absolute materialism, he must be indeed blinded by his prejudices who refuses to recognize in these analogies the evidence of intelligence and thought.

It is subsidiary to the argument from special adaptation.

I do not of course regard analogies as proofs, nor do I believe that this argument from general plan could supply the place of the great argument from design. The last lies at the basis of Natural Theology, and all the rest is merely subsidiary to the great central light. Moreover, while the argument from design comes home to every man's understanding, these analogies appeal with their full force only to the few who are able to study the processes of nature for themselves; as they alone are familiar with the phenomena in which the resemblances are seen. But to the student, whose life has been passed in successful investigation, and whose soul has been brought into sympathy with the harmonies of nature, these tokens are constantly assuring him of the presence of his God. Every discoverer feels — when in the presence of a great truth, he cannot resist the feeling — that, in discovering a law, he has been brought nearer, not to a blind agency, but to Omnipotence itself. To this conclusion he is not led solely by philosophy; for although he may defend his conviction on reasonable grounds, in its full power it transcends all human philosophy. Man cannot always tell why he knows. But when illuminated from the altar of his faith all nature wears a new aspect, and his spiritual eye discovers everywhere acting that same Infinite Intelligence which "spake in "time past unto the fathers by the prophets," and "hath in these last days spoken unto us by "his Son."

Appeals especially to scientific men.

Do I hear it said that such loose reasoning is a gross violation of the Baconian philosophy, and of that severe induction by which alone science has been built up? But do we not know, have we not seen, that the whole structure of science rests on no firmer foundation than these very analogies of nature, — that at the beginning of all knowledge, where we should most expect infallibility, we find only uncertainty and doubt?

Science is a grand temple built by man to glorify his Maker, its unfinished spire pointing to heaven, but its foundations resting on a cloud. The work has been done as well as faithful hearts and active hands could do it. Examine its walls and its buttresses, and from base-stone to coping you will find no defect. Each block has been so carefully wrought and so firmly clamped in its place, with all the strength of iron logic, that you will unhesitatingly conclude that the mighty structure has been raised, not for time, but for eternity. Yet all rests on a cloud. Let that cloud be dispersed, and only God can tell whether that structure shall stand or fall.

Are we then, you will ask, to mistrust these boasted results of science? Is this imposing structure all a phantom, a mere day-dream, from which we shall awake on the morning of eternity to find all passed? Certainly not! God has not endowed his creature with faculties of observation merely to deceive him, and with an intellect solely to lead him into error. He has not raised up the long line of scientific heroes of every age, merely to deceive themselves and mislead the world. No! the temple of science will stand fast. That cloud on which it rests is a firmer foundation than any granite rock; for it is not of man, but of God. Yet let us not forget that this assurance is based only on the same faith which is the "substance of things hoped for, "the evidence of things not seen."

"We have but faith: we cannot know;
For knowledge is of things we see;
And yet we trust it comes from Thee,
A beam in darkness: let it grow."

the kind Providence by which he has been led. And when your theory leads to this, that man has been put into a world of probation and trial and there left to walk over pitfalls with his eyes blinded, every unsophisticated mind will feel — say what you will — that the character of the God you worship is more truly symbolized by the Juggernaut car than by the cross.

I think that a great deal of false prejudice against scientific study arises from a mistaken impression that the materialist's interpretation of nature is the natural and necessary result of all scientific thought. Hence not a few religious minds have concluded that the methods of science must be all wrong, and its conclusions wholly untrustworthy. It will not, therefore, be out of place in this connection to consider briefly whether the materialist's idea of causation is the necessary, or even the probable, conclusion to which the observed facts of nature and the legitimate methods of science lead. We must remember, however, while discussing this subject, that we have passed the limits of human knowledge, and cannot therefore expect by our unaided processes of thought to prove or disprove anything. We cannot determine absolutely whether the materialist's theory be true or false; for science has not the knowledge which would enable it to form a decision. The only question for us is, whether this theory is the necessary theory, or even the most probable theory; and if it is not either the one or the other, then the theory is of no weight. One man's theory is as good as another's, provided both are equally consistent with facts. If, then, we can show on scientific grounds alone that the Christian's theory of causation is as probable as the materialist's, we shall in regard to this point also fully vindicate the religious character we have constantly claimed for scientific studies. Surely science is no more responsible for the excesses of theorists than is religion for the crimes of bigots, and it should be sufficient to satisfy any religious mind, that there is a Christian theory which is perfectly consistent with all known facts.

It is easy to understand the relative position of the two theories of causation after we have become acquainted with the facts which both must necessarily explain. Let us review, then, very briefly, these facts, which are more or less familiar to every one. An innate principle of the human mind forces upon us the conclusion that every change must have an adequate cause, and leads us to refer the phenomena of nature to what we call forces. Thus the falling of an avalanche, the flowing of the tides, the beating of the waves, the blowing of the winds, the crashing of the lightning, the burning of the fire, the moving power of steam, and the impression of light, must all have an adequate cause, and to this cause we give the name of *force*. We use this word so frequently and so familiarly that we are apt to think that we associate with it a definite conception; but a moment's reflection will show that, in regard to the nature or origin of force, we have no absolute knowledge. This

Idea of force.

word is merely our name for the unknown cause of natural phenomena. The uneducated mind naturally refers the origin of all force to the bodies from which it appears to emanate, and regards it either as a quality inherent in matter, as in the phenomena of gravitation, or as a property superimposed upon matter, as in the phenomena of light, heat, magnetism, and electricity. In either case, however, it is regarded as a *quality of matter*. Moreover, the uneducated mind, impressed most of all by the great diversity in physical phenomena, naturally infers that a similar diversity exists in the force which produces them, and thus is led to the idea that there are different kinds of force. Hence men have been led to refer the falling of bodies towards the earth to a distinct force called *gravitation*, the motion of a steam-engine to another force called *heat*, the burning of a candle to a third force called *chemical affinity*, and in like manner to each class of phenomena they have assigned a peculiar and separate force.

Language of science liable to mislead.

Such ideas as these are natural in the infancy of knowledge, and we must remember that, with all our boast of progress, the human race, so far at least as physical science is concerned, is yet in its childhood. The law of gravitation was discovered only a century and a half ago, and almost the whole of the present sciences of chemistry and physics has been developed within the lifetime of men now living. The present generation were educated in those very natural, but crude notions, and it is not until a comparatively recent period that even scientific men have been persuaded that these primitive ideas must be wholly abandoned, or at least radically modified. We are now in a transition stage, and hence arises a great difficulty in discussing the subject. The language even of modern science is based upon the old ideas, and we cannot describe natural phenomena without using terms which imply what almost all thinkers now believe to be erroneous notions. Hence, when we attempt to present spiritual views of the origin and nature of force, we are obliged to use terms which imply the opposite, and our very language appears to condemn us, or at least prejudices our theory. This is especially true of the word *force*, and we must carefully bear in mind that the origin of phenomena is not explained, because, in the language of science, they have been referred to an assumed force with a high-sounding name. Names are not things, and we know nothing more of the cause which brings the apple to the ground because Newton has called it the force of gravitation, than we did before. He gave us the law of the motion, and enabled us to predict how every apple would fall, and how every planet would move throughout space, but the cause of the motion is as closely hidden as ever. In regard to the law of gravitation we know a great deal; but in regard to the force of gravitation — whatever we may think or believe about it — we know absolutely nothing, and the same is true of every other force.

The most remarkable feature of modern science has been the constant tendency of all investigation, during the last fifty years, to show that the same energy, if only differently applied, may produce the most diversified phenomena; and now almost all the so-called forces of the old philosophy appear to be mutually convertible. Thus — to begin with a lump of coal — as we have seen, a certain amount of latent energy resides in that black mass, which has been called the force of chemical affinity. Burn the coal, — that is, combine it with oxygen, — and the affinity is satisfied, but the energy reappears as light and heat. If the coal is burnt under a steam-boiler, the heat expands the water and converts it into vapor, and then we find the energy again in the expansive force of steam. The steam expands against the piston of the locomotive, and the energy passes into the moving train. The rapidly moving mass, in forcing its way through the air and over the iron track, is constantly losing its moving power in consequence of the friction it encounters; but the energy is not lost, and if we could follow it, we should find it reappearing somewhere as heat. Suddenly the engineer opens a valve, and a portion of the energy of the steam gives motion to the air, and the effect is a shrill whistle. The brakeman applies the brakes, and the train after a few moments comes to rest. Its moving power is gone, but the energy is not lost. The motion has been transformed into heat, and the smoking brake shows where the energy has gone.

Transfer of energy.

Return now again to the lump of coal, and, instead of burning it under a steam-boiler, heat it in a properly constructed furnace in contact with roasted zinc ore. This ore is a compound of zinc and oxygen. The coal, in order to satisfy its intense affinity, seizes on the oxygen and sets the zinc free. But although the chemical affinity of the coal has been satisfied, no power has been lost; for the energy which was before latent in the carbon is now latent in the zinc. Dissolve now the zinc in dilute sulphuric acid, and the chemical affinity of the zinc will be satisfied, and, if certain conditions are fulfilled, the energy will take the form of a current of electricity. Cause this current to flow through a platinum wire, and this energy will appear in the heat and light radiated from the glowing metal. Cause the same current to flow in spiral lines around a bar of iron, and we find the energy again in the attractive force of an electro-magnet. Connect with the electro-magnet appropriate machinery, and the same energy may be so applied that it will move a light boat or turn a small lathe. Lastly, connect with the dissolving zinc four thousand miles of iron wire, and the energy will be transmitted across a continent with the velocity of thought, and write in a distant city the message which it carries.

Start next from another point. What could be more dissimilar than magnetism and light? Yet a magnet may be made the means of producing a most brilliant light. When we apply steam, or any other force, to turn a large magneto-electric machine, the moving power is transmuted by the steel magnets into a powerful electric current. If now this current is made to pass through the air between two charcoal points placed near together, the particles of coal become most intensely ignited, and emit a light which is comparable with that of the sun.* The same current, if passed through a quantity of water, will slowly decompose the liquid, and, while forcing the hydrogen apart from the oxygen, will transfer a portion of its own energy to the escaping gases, in which the power will remain latent until the elements again unite.

Illustrations like these might be multiplied indefinitely; but enough, I think, has been said to show that, to all appearance at least, the same energy may be transferred from one mass of matter to another, and that thus, while nothing but the mode of application has been changed, the power may reappear under entirely different manifestations, and produce phenomena wholly unlike those in which it was but a moment before the active cause. The truth of this principle becomes still more evident when we apply in our experiments exact measurements; for we find that in all these transfers of energy from mass to mass the power reappears undiminished. It may remain latent for a time, as in a mass of coal, but sooner or later it will reappear without having undergone the slightest loss.

* It has been proposed to use this method of illumination in some of the lighthouses on the English coast.

We must here dwell for a moment upon an important distinction, which has already been implied, between latent and active energy. It is a distinction with which every one is practically familiar, and it may therefore be made clear by referring to a few examples. A weight falling to the ground from a given height is an example of active energy, while an equal weight suspended at the same height represents an equivalent amount of latent energy. In winding up a clock, muscular energy becomes latent in the suspended weight, but reappears in mechanical motion as the clock runs down. So also a lump of coal, as already stated, represents a certain amount of latent energy. When the coal burns, its energy becomes active, and takes the form of heat. Again, there is stored up, if we may so speak, in a mass of zinc, a large amount of latent energy. If in a galvanic battery that zinc dissolves in sulphuric acid, this energy becomes active, and reappears in a current of electricity. Some persons do not like the term latent energy, and speak of energy which is not in action as potential or possible. In like manner they speak of energy in action as dynamical or actual. But terms are of no importance, if only the ideas which they express are fully understood.

Latent and active energy.

Keeping this distinction in view, we shall better understand the bearings of the important principles just before stated. When energy, in passing from one body to another, changes its mode of manifestation, it seldom flows wholly into one channel, and almost invariably more or less of it becomes latent. Thus —to go back to the example of the steam-engine— of the energy which is latent in the coal and becomes active in the form of heat when the coal burns, not more than one tenth at the most produces any useful mechanical effect. The rest becomes again latent in changing the water into steam, and in heating and expanding the iron, the bricks, the water, and the air in contact with which the fuel burns. All this heated matter represents a large amount of latent energy. It is in the condition of the wound-up weight of a clock, and as it cools, this energy is distributed to surrounding bodies. Were it now possible, at a given instant after the burning of the coal, to sum up all the energy, both active and latent, which could be traced directly back to the burning fuel, it would be found that not the smallest fraction of the energy originally latent in the mass of coal had been lost. In this case, of course, accurate experiments are out of the question; but wherever it has been possible to apply measurements, it has been found that the principle here illustrated holds true. I should not be able to make the methods of such investigations intelligible without occupying a great deal of time. Let it then be sufficient to state, that all those who have most carefully studied the subject have arrived at the same results. There is therefore every reason to believe that the principle we have been illustrating is universally true. Let us then embody it in a definite statement. *All natural phe-*

Energy never lost.

nomena are the manifestation of the same omnipresent energy, which is transferred from one portion of matter to another without any loss of power.

But if the principle as thus stated be accepted, we cannot rest here; for it involves this further conclusion, which, however marvellous, must be true. *The sum total of all the active and latent energies in the universe is constant and invariable.* In other words, power is as indestructible as matter.*

Conservation of energy.

This grand truth is generally called the law of conservation of energy, and, if it cannot as yet be regarded as absolutely verified, there can be no question that it stands on a better basis now than did the law of gravitation one hundred years ago.

But how can I give you any conception of the sublimity of the truth which this formal language implies, but which no language is adequate to express? Even poetry in the highest flights of fancy has never seen such a vision as these vistas of actual realities open to the intellect and imagination of man. Review in the light of this grand generalization the subsidiary truth which from time to time I have endeavored to illustrate during this course of Lectures,— namely, that all terrestrial energy comes from the sun. The accumulated power of the sun's delicate rays produces, as we before saw, every motion and every change which takes place on the surface of this planet, from the falling

* Many philosophers believe, with Newton, that matter in its essence is only a manifestation of power.

of an avalanche to the crawling of a worm. But that energy, as we now know, is not exhausted on the earth. To use the eloquent language of another: "Our world is a halting-place where this energy is conditioned. Here the Proteus works his spells; the selfsame essence takes a million shapes and hues, and finally dissolves into its primitive and almost formless form. The sun comes to us as heat; he quits us as heat; and between his entrance and departure the multiform powers of our globe appear. They are all special forms of solar power, — the moulds into which his strength is temporarily poured, in passing from its source through infinitude."*

Attempt now to bring together in imagination all the energies acting at one moment on the earth, and unite them in one tremendous aggregate. Begin with the moving power of the air, the hurricanes, the tornadoes, the storms, and the gentler winds which are everywhere at work from the Arctic to the Antarctic Pole, omitting in making the estimate, if you choose, the lightning and the thunder, which, though brilliant and noisy demonstrations of power, would hardly increase by a unit the vast sum. Add to this the mechanical power in the mighty flow of waters, the ocean currents, the rivers, the cataracts, the glacier-streams, and the avalanches, all over the globe. Bring into the calculation the forces at work in the various phases of

* Professor John Tyndall, in the work already quoted, "Heat considered as a Mode of Motion."

animal and vegetable life. Remember the conflagrations, the furnaces, the fires, and the other manifestations of the terrible energies of the atmospheric oxygen, whenever it is aroused. Do not even forget the comparatively insignificant power which man is wielding with the aid of powder and of steam. Making now an immense allowance for what you must have overlooked, sum this all up, — if you can without bewilderment, — and what part is it of the whole? Why, it has been calculated that it is equal to but one 2,300,000,000th of the force which the sun is every moment pouring into space. And what is the Sun? A small star in the infinitude of space, where shine Sirius and Arcturus, Regulus and Aldebaran, Procyon and Capella, with unnumbered others, all shedding forth a far mightier effluence than our feeble star; yet the grand total of the powers streaming from all the suns which human eye has seen, or which still lie undiscovered in the depths of space, alone represents the active energy of the universe. My friends, there are two theories of causation. One regards this energy as an unintelligent power. The other sees in it simply the will of the Eternal Jehovah. They are both theories. We cannot substantiate either. But which do you think is the more probable? Let us not pass hasty judgment, but soberly weigh all the testimony, and base our decision on the best scientific evidence we can obtain, and on that alone.

Thus far in our discussion we have been dealing with facts and principles which every theory of causation must explain. But we now pass into what is rather the region of speculation, and we must step more cautiously. I have used thus far the terms *energy* and *transfer of energy*, without expecting that you would attach to them any more definite meaning than that which is conveyed by the words in their most familiar use. Energy is a definite thing, which is as palpable to our senses as matter, and which, in most cases at least, we can measure as accurately. Any one who has been stunned by a blow or a fall, has been burnt by a fire, has been dazzled by the sun, or has received a shock of electricity, knows well enough what energy is; and the doctrine of the conservation of energy is wholly independent of any theory which men may entertain in regard to its essence. For this reason, I have aimed to present the grand doctrine of modern science entirely free from all speculations whatsoever; but now that we are seeking to go behind the external phenomena, it will be well for us to consider very briefly a theory which, although it does not profess to explain what energy is in its essence, nevertheless may give to the mind a more definite conception of its mode of action. The theory, it is true, cannot be regarded as fully established; but it represents the undoubted tendency of science, and the materialists would, of all others, be the first to accept it. According to the modern view, all energy appears as motion, and this too whether it be manifested in mechanical work, or in the more subtile phenomena of sound,

Energy always appears as motion.

light, heat, chemical affinity, electricity, or magnetism. We must, however, extend our idea of motion, and not limit it, as is usually done, to the motion of masses of matter.

Even the smallest material masses perceptible to our senses must be regarded as aggregates of still smaller masses, which we call molecules or atoms. These atoms, moreover, even in the densest bodies, cannot be in contact, and we must picture them to our imagination each as a tiny world, poised in space. In using the word *atom* we do not mean to imply that these particles of matter are mechanically indivisible, as its derivation would indicate, but we merely declare that they are so many separate units so far as our science is concerned. The same relation which the worlds bear to the cosmos, we conceive that these atoms bear to the micro-cosmos which every mass of matter represents, and it is believed that the motions of suns and systems have their miniature in the motions of these atoms. The ether, also, of which I spoke in the second Lecture as filling celestial space, is supposed to pervade equally the atomic spaces, to surround each atom with a highly elastic atmosphere, and to be the medium by which motion is transmitted throughout a universe which includes the infinitesimal as well as the infinite. Moreover, we conceive that the motion of the atom is the exact counterpart of the motion of a world or of the motion of a ball, and that all motion obeys the selfsame laws. As when

Motion of atoms to be considered.

an ivory billiard-ball strikes another, it gives up the whole or a part of its motion to the second ball, so we believe that one atom may transmit motion to another. In like manner as an impulse is transmitted through a long line of billiard-balls, and the last ball only appears to move, so also we conceive that the electrical impulse is transmitted from atom to atom through the telegraph wire, and produces perceptible motion only when transformed into magnetism at the end of a thousand miles. Again, motion may be transmitted from atoms to masses of matter; for although the impulse imparted by a single atom may be as nothing, the accumulated effect of millions on millions of these impulses may be immense. In this way, we conceive, the motions of the ether particles in the sunbeams unite to produce all the grand phenomena of nature. On the other hand, the motion of great masses may be suddenly resolved into the motions of the atoms composing these masses, and thus, when motion outwardly appears to cease, it may only be transferred from the previously moving body to the atoms within. When the cannon-balls, with their immense velocity, strike the Ironsides, and fall harmlessly from her armor-plates, the particles of iron take up the motion of the ball, and indicate by a higher temperature that the energy has not been lost.

Understanding then the term *motion* in the extended sense just explained, we shall comprehend more clearly the theory stated above. This theory supposes that the phenomena of sound, light, heat, and electricity are produced by the motions of atoms, in the same way that the grander phenomena of mechanics and astronomy are caused by the motion of large masses of matter. The transmission of energy is, then, the direct result of the transmission of motion, and the conservation of energy is fully explained by the well-known law of inertia, which the motions of all matter necessarily obey. I have not time to enter into any details in regard to the mode of motion by which light, heat, and all this class of phenomena are produced, other than those already given in the previous Lectures of this course; but I take great pleasure in referring my audience to the work of Professor Tyndall, already frequently quoted, which is by far the best popular statement of the subject that has ever been made. Indeed, great differences of opinion in regard to the mode of the atomic motion are entertained by those who accept the theory in its general statement, and in many cases we can form no conception of the peculiar phase which the motion assumes. It is sufficient for my purpose if I have been able to make clear the general principle, and I will only add a few numerical results, which will show what a precise form the theory has taken in the minds of scientific men.

According to the modern theory, when we heat a body we merely impart to its atoms a greater velocity of motion. Now, according to the experiments of Professor Joule, when we raise the temperature of a pound of water two Fahrenheit degrees, we distribute among the atoms of the liquid an amount of motion equal to that acquired by a weight of two pounds in falling 772 feet; and a simple calculation will show that this is represented by a Minie ball, weighing one eighteenth of a pound, moving with a velocity of 1,338 feet in a second.* The amount of motion, therefore, which is imparted to the particles of water in an ordinary tea-kettle during the process of boiling, must be in the aggregate vastly greater than that ever acquired by any projectile. We shall arrive at a still more remarkable result, if we examine in the light of our theory the process of chemical combination by which water is formed. In this process of burning, one pound of hydrogen gas combines with eight pounds of oxygen gas to form nine pounds of water. Although the distances which separate the atoms of the two gases before combination are utterly inappreciable by our senses, yet, in passing over these distances, they acquire a velocity which causes them to clash together with tremendous energy, and in the collision this form of atomic motion is transmuted into that other mode of motion which we call heat. Incredible as it may appear, the amount of motion which in the act of combination alone is thus transmuted into heat corresponds to the fall of a ton weight down a precipice 22,320 feet high. Such illustrations might be multiplied indefinitely; but you will see from

Amount of the atomic motion.

* In making the calculation, it must be remembered that the amount of motion is measured by the square of the velocity.

these how purely mechanical the idea is which we associate with the motion of an atom, and you must have been impressed by the magnitude of the energy which these atomic motions represent. "I have seen," says Professor Tyndall, "the wild stone avalanches of the Alps, which smoke and thunder down the declivities with a vehemence almost sufficient to stun the observer. I have also seen snow-flakes descending so softly as not to hurt the fragile spangles of which they were composed; yet to produce from aqueous vapor a quantity of that tender material which a child could carry, demands an exertion of energy competent to gather up the shattered blocks of the largest stone avalanche I have ever seen, and pitch them to twice the height from which they fell." If such, then, be the measure of these atomic motions, we can easily conceive how the motion of the cannon-ball might be transferred to the particles of the armor-plate without much apparent result, and even how the energy of a world might be maintained by the motion of the atoms in the sunbeam.

Accepting, then, this new theory of science, and admitting that all energy is manifested in motion, we reduce at once our discussion of the doctrine of causation to this simple question, — What is the primary cause of motion? If we can explain the simplest case of motion, we have solved the problem for the universe. Take, for example, a boy's ball, moving through the air under the impulse of a well-directed blow. Do we not know something of the cause of that motion? Is it not connected with the muscular contraction of the boy's arm, produced by his will? Is not his volition, acting mysteriously on matter, at least the occasion of the motion? It is perfectly true that the will does not create the motion. The ball is impelled by a portion of that energy in nature which man can neither increase nor diminish. But still the boy's will is the occasion of the motion. It has opened the channel through which the energy of nature has flowed to produce the specific result which the boy desired. So, in a thousand other ways, man is able to come down, as it were, upon nature, and to introduce a new condition into the chain of causation. Place the point of contact as far back as you please, theorize about the subject as you may, the fact still remains the same. Our will does act on matter, and does act to produce most efficient results. Here is energy exerted of whose cause we have the consciousness within ourselves, and, if the analogy is worth anything, it points to but one conclusion, — namely, that motion is always the manifestation of will. As the boy's will acted on that atom of matter, which, though moved perhaps but a hair's breadth from its position, set in action — as if by the touching of a spring — the train of natural causes which gave motion to the ball, so we may suppose that the Divine will acts in nature. According to this view, the energy which sustains the universe is the will of God, and the law of conservation is only the manifestation of His immutable being, — "the same yesterday, and to-day, and forever."

We do not say that this theory can be proved, for certainty here is out of the question; but we do claim that it is based on the only analogy which nature affords, that it is a legitimate deduction of science, and that it is perfectly consistent with Christian faith. On a subject where science can only grope, the wildest theories are possible; but these should not trouble a well-balanced mind, so long as there exists an equally probable theory which can be reconciled with the purest faith. It has been my aim in this Lecture to show, not only that such a theory is tenable, but also that the Christian theory of causation is the most probable theory of science; and my earnest hope is, that, for some minds at least, the considerations I have offered will help to reconcile the apparent conflict between science and religion, which materialism is ever striving to foment. Allow me to add, in concluding, one or two other suggestions which may be of value in the same direction.

Necessary limitations of scientific and religious thought.

I cannot but believe that the appearance of clashing between science and religion would be wholly avoided, if the teachers both of God's unwritten and of His written word would pay more regard to the necessary limitations of scientific and religious thought. On subjects where the methods of acquiring knowledge are so utterly unlike, where the relations of the knowledge to the human understanding are so different when gained, it is in vain to expect literal accordance. Science, both in its methods and its results, addresses the understanding exclusively; Christianity appeals chiefly to the heart. Science aims to instruct; Christianity aims to persuade. Science is attained by study, and is possible only for the few; Christianity is a free gift from God to all men who will receive His Son. The results of science are fully comprehended, and can be expressed in definite terms; the truths of Christianity stand on a level above man's intellect, and can only be shadowed forth in types and symbols. The forms of science are constantly changing; the types and symbols of Christianity are permanent. Lastly, while the language of science may be so varied from time to time as to express accurately the current ideas, Christianity necessarily retains the forms through which it was first revealed. Under such conditions how can it be expected that the language of revelation should agree with the letter of science? You might as reasonably find fault with nature because its crystals are not perfect, as criticise the Bible because its language, although embodying divine truth, is not free from the necessary limitations and imperfections of the human medium of thought.

I have not time to develop the ideas here suggested; but the bearing of the facts is obvious, and they lead to important practical conclusions. In the first place, they should

Practical conclusions.

teach men of science to honor and reverence the forms of religion. They are the types and symbols of a higher truth than any Science can teach. Let Science vindicate her own methods, and allow no interference within her proper sphere; but unless she learn that there are other sources of knowledge than material nature, and other channels of truth than the intellect, her own philosophy will be confounded, and her light will go out in darkness. On the other hand, it is equally the duty of the ministers of religion to honor and respect the methods of science. They have been ordained by God, and through these processes of thought He is constantly revealing eternal truths to the mind of man. Insist as strongly as you please that Science should be allowed no voice in matters of faith. Scrutinize as closely as you can every step of her logic; but so long as she keeps within her legitimate province, allow her the largest liberty, and extend to her the most generous encouragement. Watch sharply her results, and expose her fallacies wherever you can find them; but, if your judgment condemns, let it be on scientific grounds, and not by any arbitrary standard of your own. Above all, even if you think your most cherished opinions are in danger, do not withdraw your fellowship hastily, or be betrayed into undiscriminating censure. Science is paramount within her own province. Do everything in your power to consecrate her aims and sanctify her spirit, but do not attempt to control her investigations or restrict her free thought. Await God's time. If science is wrong, she will sooner or later correct her error. If she is right, God is on her side, and you would find yourself fighting against the great Jehovah.

Again, a proper appreciation of the necessary limits of scientific and religious thought should lead all men to reverence the Word of God as it has been handed down to us through history. In view of the facts already intimated, I cannot look with favor on any attempts at Biblical criticism which aim to square the language of Scripture by the results of modern science. They leave a most unpleasant impression on my mind. I believe the Bible inspired, from the grand epic of creation, with which it opens, to the glorious vision of the New Jerusalem at its close. I feel that its very words are sacred, and if you are so ready to accommodate any part of them to the shifting phases of science, what certainty can I have in regard to the whole? The Bible is no text-book of science, and the attempt to impose an equivocal or mysterious meaning on its simple and obvious statements degrades and dishonors it in the minds of devout men. The methods by which its truths are expressed may be at times rough and uncouth; but they are the methods chosen by God, consecrated by the blood of martyrs, and hallowed by the tears of saints; and they have therefore a power which no other language could have. Break not the mould in which the forms of faith have been cast, before they have become firm and hard, lest the precious metal should itself be lost. Finally, leave religion and science to their respective methods, and encourage both alike in their noble calling. Let science, by cultivating man's intellect, elevate him to nobler and more spiritual views of God's wisdom and power. Let religion, by purifying man's heart, open to him clearer visions of God's purity and love; and, at last, when this material shall have vanished, and when the waters of controversy shall have ceased to roll, the heart and the intellect, made one and washed clean in the blood of the Lamb, shall unite in the song of the angels around the throne, saying, "Blessing, and glory, and wis-"dom, and thanksgiving, and honor, and power, "and might, be unto our God for ever and ever."

Limits of natural theology.

But while insisting upon the necessary limitations of scientific and religious thought, I must not forget that all such considerations bear with peculiar force upon the questions which it has been my privilege to discuss in this course of Lectures. Therefore, although I have most carefully endeavored to guard my argument from the slightest exaggeration, I should not feel justified in concluding without distinctly stating how far in my opinion the argument of natural theology may be safely carried, and to what extent unaided science may be said really to prove the fundamental truths of Christianity.

In the first place, then, I believe that the existence of an intelligent author of nature, infinite in wisdom and absolute in power, may be proved from the phenomena of the material world with as much certainty as can any general truth of science. In the second place, I am of opinion that the facts of nature also prove — although the arguments adduced may be less convincing — that the author of nature is a personal being, and the one only and true God revealed to us in the Bible. Lastly, I think that the relations of the human mind to the material world, viewed in the light of modern science, give us strong reason to believe, on scientific grounds alone, that the universe is still sustained, in all its parts, by the same omnipotent and omniscient Will which first called it into being. It may be that God intrusts the execution of His will to subsidiary intelligences, and such a belief is perfectly consistent with the above conclusions. But nature throws no light upon the subject, and therefore I have not discussed this question. If, however, such agents exist, the facts of nature just referred to would tend to show that they must be intelligent causes, and subject to Him who is all in all.

To the extent I have indicated I regard the argument of natural theology as logically valid. Moreover, I am persuaded that science confirms and illustrates the priceless truth which Christ came on earth to reveal; but I do not believe that the unaided intellect of man could ever have attained to certainty in regard to even the least of these truths, independently of revelation. Nevertheless, as I stated in my introductory Lecture, I feel that the best service which science can render to religion is in the way of confirmation and illustration, rather than in that of absolute proof, and for this reason I have preferred to discuss my subject chiefly from that point of view.

The subject, as prescribed by the founder of these Lectures, is "The power, wisdom, and goodness of God as manifested in His works," and to this form of statement, if interpreted in the sense just indicated, I have nothing to object. I do not believe, however, in any sense, that nature proves the goodness of God. When the heart has been once touched by the love of God, as manifested on Calvary, the tokens of God's goodness are seen everywhere: but before this, nature, to one who has seen its terrors and felt its power, looks dark indeed; and the pretence that the material universe, unexplained by revelation, manifests a God of unmixed beneficence, not only does harm to religion, but places science in a false light. The most superficial observation shows that this is not true. Lightning and tempest, plague, pestilence, and famine, with all their awful accompaniments, are no less facts of nature than the golden sunset, the summer's breeze, and the ripening harvest; and who does not "know "that the whole creation groaneth and travaileth in "pain together until now"? It does not change the terrible fact to say that nature has been disordered by man's sin; for sin is itself the greatest evil in the world, and its ghastly forms meet us at every

Nature does not prove God's beneficence.

step. So prominent, indeed, is the evil in nature, and so insidiously and mysteriously does it pervade the whole system, that an argument to prove the malignity of God could be made to appear quite as plausible as the arguments which are frequently urged to prove His pure beneficence; and wherever the unaided human intellect has attempted to make to itself a beneficent God, it has always made a malignant Deity as well. Nature manifests God's wrath no less than His love, and it is a false and sickly philosophy which attempts to keep the awful fact out of sight. God is our Father; but nature could not teach it, and "the Word was made flesh" to declare it. God is love; but nature could not prove it, and the Lamb was "slain from the foundation of the world" to attest it. Nature is but a part of God's system, and not until the natural and the supernatural shall be made one will the mystery of evil be solved.

THE END.

Cambridge: Stereotyped and Printed by Welch, Bigelow, & Co.